Design Thinking

Design Thinking is a set of strategic and creative processes and principles used in the planning and creation of products and solutions to human-centered design problems.

With design and innovation being two key driving principles, this series focuses on, but is not limited to, the following areas and topics:

- User Interface (UI) and User Experience (UX) Design
- Psychology of Design
- Human-Computer Interaction (HCI)
- Ergonomic Design
- Product Development and Management
- Virtual and Mixed Reality (VR/XR)
- User-Centered Built Environments and Smart Homes
- Accessibility, Sustainability and Environmental Design
- Learning and Instructional Design
- Strategy and best practices

This series publishes books aimed at designers, developers, storytellers and problem-solvers in industry to help them understand current developments and best practices at the cutting edge of creativity, to invent new paradigms and solutions, and challenge Creatives to push boundaries to design bigger and better than before.

More information about this series at https://link.springer.com/bookseries/15933

Designing Our Future

Redefining Design in the Age of AI for Sustainable Innovation

Diana Olynick

apress®

Designing Our Future: Redefining Design in the Age of AI for Sustainable Innovation

Diana Olynick
Winnipeg, MB, Canada

ISBN-13 (pbk): 979-8-8688-1162-3 ISBN-13 (electronic): 979-8-8688-1163-0
https://doi.org/10.1007/979-8-8688-1163-0

Copyright © 2026 by Diana Olynick

This work is subject to copyright. All rights are reserved by the Publisher, whether the whole or part of the material is concerned, specifically the rights of translation, reprinting, reuse of illustrations, recitation, broadcasting, reproduction on microfilms or in any other physical way, and transmission or information storage and retrieval, electronic adaptation, computer software, or by similar or dissimilar methodology now known or hereafter developed.

Trademarked names, logos, and images may appear in this book. Rather than use a trademark symbol with every occurrence of a trademarked name, logo, or image we use the names, logos, and images only in an editorial fashion and to the benefit of the trademark owner, with no intention of infringement of the trademark.

The use in this publication of trade names, trademarks, service marks, and similar terms, even if they are not identified as such, is not to be taken as an expression of opinion as to whether or not they are subject to proprietary rights.

While the advice and information in this book are believed to be true and accurate at the date of publication, neither the authors nor the editors nor the publisher can accept any legal responsibility for any errors or omissions that may be made. The publisher makes no warranty, express or implied, with respect to the material contained herein.

Managing Director, Apress Media LLC: Welmoed Spahr
Acquisitions Editor: James Robinson- Prior
Desk Editor: James Markham
Editorial Project Manager: Gryffin Winkler

Cover designed by eStudioCalamar

Distributed to the book trade worldwide by Springer Science+Business Media New York, 1 New York Plaza, New York, NY 10004. Phone 1-800-SPRINGER, fax (201) 348-4505, e-mail orders-ny@springer-sbm.com, or visit www.springeronline.com. Apress Media, LLC is a Delaware LLC and the sole member (owner) is Springer Science + Business Media Finance Inc (SSBM Finance Inc). SSBM Finance Inc is a **Delaware** corporation.

For information on translations, please e-mail booktranslations@springernature.com; for reprint, paperback, or audio rights, please e-mail bookpermissions@springernature.com.

Apress titles may be purchased in bulk for academic, corporate, or promotional use. eBook versions and licenses are also available for most titles. For more information, reference our Print and eBook Bulk Sales web page at http://www.apress.com/bulk-sales.

Any source code or other supplementary material referenced by the author in this book is available to readers on GitHub. For more detailed information, please visit https://www.apress.com/gp/services/source-code.

If disposing of this product, please recycle the paper

Table of Contents

About the Author xiii

About the Technical Reviewer xv

Acknowledgments xvii

Preface xix

Introduction—The Challenge of Futures Design and AI xxv

Seeing Further, Acting Sooner xxix

Part I: The Foundations of Design and Futures Thinking 1

Chapter 1: The History of Design and Contemporary Design Thinking ... 3

   1.1 Evolution of Design Principles and Practices 3

   1.2 Key Movements and Figures in Design History 5

      1.2.1 The Arts and Crafts Movement 6

      1.2.2 Art Nouveau 6

      1.2.3 The Bauhaus Movement 6

      1.2.4 Modernism 7

      1.2.5 Art Deco 7

      1.2.6 Mid-Century Modern 7

      1.2.7 Postmodernism 8

      1.2.8 Sustainable and Contemporary Design 8

   1.3 Contemporary Design Thinking and Futures Thinking 9

   1.4 Summary 11

TABLE OF CONTENTS

Chapter 2: The History of Futures Thinking 13
2.1 Origins and Evolution of Futures Thinking 13
2.2 Major Theorists and Approaches in Futures Studies....................................... 15
2.3 Integration of Futures Thinking in Modern Disciplines 17
2.4 Summary... 20

Part II: Exploring Futures Design 21

Chapter 3: Core Concepts in Futures Thinking 23
3.1 Defining Futures Thinking: Beyond Traditional Practices 23
3.2 The Emergence of Scenario Planning and Speculative Design..................................... 25
    3.2.1 Scenario Planning.. 26
    3.2.2 Speculative Design.. 27
3.3 The Evolving Role of Strategic Foresight in Design... 29
3.4 Summary... 31

Chapter 4: Global Perspectives and Cultural Impact 33
4.1 Diverse Cultural Viewpoints in Futures Thinking... 33
4.2 Designing for a Multifaceted World... 35
4.3 Case Studies: Global Futures Design 37
4.4 Summary... 42

Chapter 5: Ethical and Sustainable Design... 43
5.1 Navigating Ethical Complexity in Futures Thinking.. 43
5.2 Sustainability As a Core Principle 45
5.3 Long-Term Impacts and Responsibilities.. 48
5.4 Summary... 50

TABLE OF CONTENTS

Part III: Integrated Futures Thinking Methodologies 51

Chapter 6: Foundational Methodologies in Futures Thinking 53

6.1 Introduction to Futures Thinking Concepts and Approaches........................ 53

    6.1.1 Overview of Futures Design and Speculative Design Basics 57

    6.1.2 Understanding the Future: Definitions, Theories, and Influences 59

6.2 Framework for Anticipating Futures 61

    6.2.1 Introduction to the STEEP+V Model and the Futures Cone 63

    6.2.2 Identifying and Tracing Signals of the Future 68

    6.2.3 The Role of Speculative Design in Shaping Futures Scenarios 70

6.3 Summary.. 73

Chapter 7: Tools and Techniques for Futures Thinking Exploration....75

7.1 Scenario Development and Analysis... 75

    7.1.1 Creating and Analyzing Scenarios Using the Futures Wheel and Futures Cone 81

    7.1.2 Applying Jim Dator's Four Futures Framework 87

    7.1.3 Building Detailed Future Scenarios and World Building Techniques 89

7.2 Imagining and Prototyping Futures 93

    7.2.1 Radical Reimagining and Innovation for Desirable Futures.................. 95

    7.2.2 Developing Future Personas, Locations, and Artifacts 98

    7.2.3 Prototyping Future Concepts and Models... 102

7.3 Summary.. 104

Chapter 8: Strategic Foresight and Backcasting 107

8.1 Strategic Foresight Principles and Approaches 109

    8.1.1 Application in Market and Technology Trends 111

    8.1.2 Creating a Strategic Futures Playbook for Brands and Products 113

8.2 Backcasting: Charting the Path to Desired Futures 115

    8.2.1 Understanding and Applying the Backcasting Methodology 116

8.2.2 Roadmapping and Root Cause Analysis 119
8.2.3 Backcasting Actionable Roadmaps from Preferred Futures 121
8.3 Summary 124

Chapter 9: Synthesizing Futures Practice 127

9.1 Contextualizing Futures in Place, Culture, and Community 130
    9.1.1 Role of Place in Futures Design 132
    9.1.2 Tools and Frameworks for Capturing Insights and Assessing Impacts 134

9.2 Final Project: Comprehensive Futures Project 136
    9.2.1 Selecting a Focal Issue/Domain 136
    9.2.2 Applying Futures Methodologies 137
    9.2.3 Presenting and Peer Reviewing 139
    9.2.4 Selection of a Focal Area for Futures Design 139
    9.2.5 Application of Methodologies Learned: Scenario Building, Prototyping, Strategic Foresight, and Backcasting 142

9.3 Case Study: The Future of Roads and Highways 188

9.4 Summary 245

Part IV: AI in Futures Thinking: Opportunities and Challenges 247

Chapter 10: AI and Futures Thinking Methodologies 249

10.1 AI in Scenario Development and Analysis 250
10.2 Case Studies: AI-Driven Futures Thinking and Speculative Design 253
10.3 Ethical Implications of AI in Design 256
10.4 Summary 258

TABLE OF CONTENTS

Part V: Advanced Concepts and Methodologies 261

Chapter 11: Interactive and Transdisciplinary Approaches with AI..... 263

11.1 AI in Interactive and Digital Media Design 265

11.2 AI's Role in Blending Design with Other Disciplines 268

11.3 Interviews with AI and Design Industry Leaders 271

11.4 Summary... 290

Chapter 12: Futures Thinking in Practice with AI 293

12.1 Practical Application of AI in Futures Thinking 294

12.2 AI-Enhanced Workshop Models and Projects 300

12.3 AI in Futures Thinking for Policy Making 304

12.4 Summary... 306

Part VI: Extending the Boundaries 309

Chapter 13: Emerging Fields and New Frontiers in AI 311

13.1 AI's Role in Futures Thinking for Space Exploration 312

13.2 AI in Post-human and Virtual Reality Design 313

13.3 AI and Futuristic Infrastructure in Design 315

13.4 Summary... 317

Chapter 14: AI in Complex Environmental Futures Thinking 319

14.1 AI Strategies for Climate Adaptation Design 320

14.2 AI in Design for Post-scarcity Economies 322

14.3 AI, Neurodiversity, and Inclusion in Design 324

14.4 Summary... 327

Chapter 15: Philosophical and Mythological Perspectives on AI and Futures Thinking 329

15.1 AI, Mythology, and Philosophy in Futures Thinking 329

15.2 AI's Influence on Futures Thinking Mental Models 330

15.3 AI in Transdisciplinary Design: Merging Thought with Practice 332

15.4 Summary 333

Part VII: Societal Systems in the Age of AI and Futures Thinking 335

Chapter 16: Societal Systems in the Age of AI and Futures Thinking 337

16.1 AI in Designing Future Health Ecosystems 338

16.2 AI's Role in Future Law, Justice, and Political Systems 341

16.3 AI in Media and Information Systems Design 345

16.4 AI in Environmental and Ecological Systems Design 348

16.5 AI's Impact on Cultural and Social Systems Design 350

16.6 AI in the Future of Financial and Banking Systems 353

16.7 AI in Future Transportation and Infrastructure 355

16.8 AI in Agricultural and Food Systems Design 358

16.9 AI and the Future of Family and Relationships Systems 360

16.10 Summary 362

Part VIII: Concluding Perspectives 365

Chapter 17: Futures Thinking: Looking Ahead in the Age of AI 367

17.1 The Future of Futures Thinking with AI: Reflections and Predictions 368

17.2 AI's Role in Addressing Design Challenges and Opportunities 371

17.3 Final Thoughts: AI's Place in the Path Forward in Futures Thinking 372

TABLE OF CONTENTS

Appendix A: Tool Box for Futures Thinking..........................375

Appendix B: Glossary of Futures Thinking..........................391

Appendix C: Additional Resources for Further Learning...................395

Appendix D: References..401

Index..409

About the Author

Diana Olynick is a leader in engineering innovation, emerging technologies, strategic foresight, and futures design. As a registered Professional Engineer, Diana leverages her extensive experience in the public and private consulting sectors to drive sustainable and forward-thinking solutions. Her expertise includes AI-driven strategic foresight methodologies and the application of cutting-edge technologies in engineering. Beside her technical engineering experience, Diana has been a thought leader in spatial computing and conscious design, guiding international audiences through the complexities of the immersive and AI-enhanced future. Diana is also the author of *Interfaceless,* another work dedicated to reshaping the future of design and technology through sustainable innovation. Diana's educational contributions can be found at https://www.dianaolynick.com.

About the Technical Reviewer

Kamila Iżykowicz is a futures designer, researcher, and internationally exhibited artist (MA Design Futures, RCA). She leads Futures Design (www.futuresdesign.co.uk), a practice blending research, storytelling, and design to explore how sensory experience shapes time, legacy, and social–ecological systems. Her projects have been showcased at Tate Modern, Dubai Futures Forum, Design Indaba, and major design weeks across Europe. Her awards include CERN's Far Futuring Grand Challenge and the Creative Conscience Prize. She has published in academic journals and presented at institutions including Kingston University, Northumbria, the University at Buffalo, and the Royal College of Art, as well as with industry partners such as John Lewis Partnership.

Acknowledgments

First and foremost, my deepest gratitude goes to my family—Sara, my mom, Stephen, Gani, and Penny—whose unwavering belief in this work anchored me through the busiest and most challenging moments of the journey.

To the diligent teams who helped bring this vision to life: Welmoed Spahr, James Robinson-Prior, James Markham, Gryffin Winkler, and Shobana Srinivasan. To our technical reviewer, Kamila Iżykowicz, and to our world-class contributors—Jerome Glenn, John Maeda, Alice Rawsthorn, and Nabil Harfoush—thank you. To every editor, manager, colleague, and friend who offered guidance and critique along the way, your contributions were and continue to be invaluable.

As you are about to discover through *Designing Our Future*, it is all about the now. I invite you to pause for a brief meditation on the present moment, not only here, but throughout these explorations and in your own practice. It can feel contradictory to pick up a book about the future when we are so often told to stay present. And yet that is precisely the point: from grounded presence we can build a more resilient and conscious tomorrow. That, ultimately, is why this book exists.

Most importantly, dear reader, Thank You. By engaging with these pages, you make the work alive. I hope, in some small way, it equips and encourages you to shape the resilient and innovative futures ahead.

Preface

"The future is in your hands."

Imagine you stumbled across a strange device in one of your trips to the Middle East. It is a watch. You take it with you and begin to study it trying to decipher its symbols. As you touch the top button, you start to feel dizzy; you feel you are about to faint. Somehow you open your eyes and you are in another place, completely different from your last hotel room where you had been inspecting the artifact. You look through the window and you can see something is just not matching your previous reality.

The once vibrant city life has been replaced by unsettling silence. You can hear the wind through the remains of skyscrapers, their facades being replaced by decades of neglect and contamination. Nature, once pushed back by the relentless treatment of humanity, now reclaims the abandoned streets. Cracked asphalt is fractured by blades of grass.

Inside a decaying apartment, sunlight filters through a broken window, and as you observe, years of unchecked climate change, driven by industrial expansion and resource extraction, have wreaked havoc. The global climate system is spiraling out of control, and the window for course correction has slammed shut. Food production is projected to collapse within the next year. Mass starvation and social unrest are inevitable.

As you witness the chaos, you remember stories from your grandparents before the environmental crisis. Images project in your mind: fresh green fields, prosperous food markets, people happy and smiling... Now, those memories feel like a cruel dream.

PREFACE

Deep down you know the truth. Humanity gambled on the future, prioritizing short-term gain over long-term sustainability. Even in this desolate world, there must be a way forward, you think. You decide to work on what seems now a dream—the dream of rebuilding, not just a society, but a future where humanity learns from its mistakes. Together, with the few surviving communities, you might find a way to restore balance and hope in a world ravaged by human excess and neglect. The year was 2125.

Suddenly you wake up in your own bed, back in time. In our current world.

This scenario serves as a stark reminder of the consequences of inaction. It underscores how the choices we make today, both in design and technology, shape the trajectory of our future.

This book explores the critical intersection of design, artificial intelligence, and the urgent need for sustainable, long-term thinking. The world is undergoing unprecedented technological advancement, fundamentally altering the way we design products, services, and systems. Yet, the traditional design mindset struggles to adapt. It's inherently reactive, limited by its focus on immediate problems rather than the interconnected systems of our increasingly complex world.

Integrating Futures Thinking with the power of AI is crucial. Futures Thinking provides a framework to anticipate challenges, explore potential scenarios, and design proactively for a better future. AI offers enormous potential as a tool for creative exploration, pattern recognition, and strategic foresight. Together, they provide an invaluable toolkit for shaping the world we want to live in.

This book is intended for designers, technologists, policymakers, and anyone interested in actively designing the future, "the forward thinking ones." It will guide you through understanding Futures Thinking principles and methodologies, integrating AI tools strategically, and applying these frameworks to real-world challenges.

PREFACE

This book begins by defining Futures Thinking and outlining its key principles. It explores the ways AI transforms design processes, and highlights the ethical considerations of this powerful partnership. The remainder of the book dives deeper into practical methodologies, providing concrete tools and examples to help you implement these ideas into your work.

Before we dive deeper into the content of each section, this is where we need to make the most important clarification. If you have been studying these fields, you might have heard the terms futures thinking, futures design, strategic design, critical design, speculative design, anticipatory design, futures studies, and more. This book will focus on the broad span of the methodologies used in strategic foresight, futures thinking, and futures design.

It is beneficial to notice that the process of exploring signals of change, determining preferable futures, and creating scenarios often touches on different aspects of foresight, futures thinking, and futures design, frequently overlapping and complementing each other. In fact, I have seen different experts talking about different aspects of these disciplines interchangeably.

Due to their interrelated nature, these practices are not rigidly separated. For example, when you are exploring signals (foresight), you might simultaneously begin to imagine how those signals could manifest (futures thinking), and as you develop scenarios, you might decide to create visual or experiential representations (futures design) or push the boundaries with more radical, thought-provoking concepts (speculative design).

And finally, the most important takeaway is that throughout the world, there are an important number of researchers, futurists, and academics that have dedicated their entire careers to this field, and even with that depth of practice, their opinions about the use of these different methodologies and their definitions can also vary to some degree, even when all fall under the same unified study of the future. In fact, some of the design thinking methodologies have been morphed and adjusted for the study of the future.

PREFACE

Ultimately what is important is that you study the methodologies, through your own experience learn how they are best applied in your respective projects or fields, and regardless of their names or characterizations, you are able to produce solid insights about potential futures for sustainable development and social good. I even invite you to create your own methodologies once you understand the principles so that you can customize your own tool for the specific needs of your job.

Lastly, here is my main contribution for you to remember about this book. I would like to share this through the actual story of how this unfolded. I have since the 2010s followed and learned from the work of a researcher that I appreciate for the contributions he has made. His name is Dr. Jerome Glenn. One day we were talking through a video call and the point he impressed on me was that foresight is only as strong as its evidence chain. You can apply a method correctly and still struggle to explain how signals became insights, or why a scenario was judged "preferable." From that conversation forward, I have tried to make the logic from Signal ➤ Insight ➤ Implication ➤ Choice visible and auditable.

This is an important warning for us to keep in mind, on how much we really know about where conclusions come from and how much data we really have gathered and rigorous research methodologies followed to arrive at them. Please keep in mind that this book is not about the process to establish rigorous research alone, as that would be a completely separate scope, but to be aware that this might be required for a complete and formal futures study. In this book we will work on understanding and reflecting on the methodologies, and brief case studies will be provided using assisting tools like AI, but the complete results of a formal futures study would require a strong infrastructure of experts teams working together along with associated communities, stakeholders, and participatory frameworks that can lead to those objective insights. The content presented in this book is presented for methodological purposes and the examples presented are for the purposes of understanding the tool, not to provide the ultimate conclusion of the data used for the examples.

My own journey in the facets as an engineer, researcher, designer, and technologist has profoundly shaped my perspective. Having seen both the potential and the pitfalls of unchecked technological advancement, I'm firmly convinced that intentional design is critical.

It is my sincere hope that this book empowers you to embrace Futures Thinking and harness AI responsibly. The future isn't pre-determined. It's an act of continuous design, and we get to be the architects.

Introduction—The Challenge of Futures Design and AI

What world are we creating?

In an age of rapid technological change, interconnected global systems, and looming environmental crises, one thing is certain—the future will be complex. Reactive design approaches that focus on solving today's problems are woefully insufficient. Building a world that is sustainable, equitable, and resilient requires a fundamental shift in how we think about design itself.

This shift is embodied in Strategic Foresight. Rather than predicting one outcome, we explore a portfolio of futures: the possible, plausible, and probable, and then clarify the preferable trajectories we aim to shape. Using this wider lens helps us anticipate long-term impacts of today's design choices and align action with ethical and sustainable intent.

Artificial intelligence (AI) has become an indispensable tool in this endeavor. AI's capacity to analyze data, identify patterns, and generate creative options expands the designer's toolkit. It allows us to navigate the complexities of the future and make more informed, far-reaching decisions. However, this powerful technology also demands a sense of responsibility. As we design alongside AI, ethical considerations become paramount.

This book aims to be your guide to this exciting intersection of Futures Thinking, AI, and the creation of a sustainable future. Within these chapters, you will explore how design can be used to address the most pressing challenges of our time, from climate change to social inequality. You'll discover methods for harnessing AI in design without losing sight of the human experience. Most importantly, you'll gain the tools to become an active participant in shaping a future that benefits all.

INTRODUCTION—THE CHALLENGE OF FUTURES DESIGN AND AI

In Part I, we set out to understand the ground bases at the intersection of design and futures thinking, exploring the different approaches being tried in both fields and the integration of them in modern disciplines like cognitive science, computational design, bioscience, social studies, and more. This understanding of the traditional approach to design, at the contemporary application reflected in futures design, led us to reflect on Part II about the foundations of the core concepts in futures design thinking, as well as its cultural and global impact over the last years along with its ethical and sustainability viewpoint and direct applications. Considering the variable approaches in futures design, in Part III we discuss different methodologies and frameworks as well as tools and applications to gain a better understanding of what devices are better used in different contexts. Since different fields offer different challenges, Part IV is dedicated to explore some of those ones as well as opportunities in emerging fields. Part V then takes all these building blocks and presents the use of AI resources in the field of Futures Design while deepening its applications with some of the emerging fields discussed in Part IV, along with complex environmental scenarios and the different philosophical and mythological perspectives. Part VI explorers how AI pushes futures thinking beyond the typical constraints opening new frontiers in space exploration, post-human and virtual realities, regenerative environmental systems and inclusive design among others. In Part VII, we examine societal systems such as politics, health, economics, education, environment and more, concluding with Part VIII as a final reflection on opportunities moving forward and anticipations on the integration of new methodologies and frameworks using advanced technologies. Designers and leaders make better choices when they can imagine desirable futures and work backward to today's decisions. This book combines strategic foresight (evidence, scenarios, options) with futures design (experiments, artifacts, engagement) so teams can (1) define what preferable means for them, (2) explore multiple trajectories beyond straight-line trends, and (3) translate vision into near-term moves they can test and measure. See the section "Radical Reimagining and Innovation for Desirable Futures" in Chapter 7 for the mindset shift that powers the tools that follow.

The future is not something that simply happens to us. It's a canvas awaiting our design. Let's start building.

Seeing Further, Acting Sooner

We are making decisions today that will live for decades, often under conditions that shift faster than our plans. This book is about developing the practical habits to see further and act sooner. At its core is strategic foresight: clarifying purpose, scanning widely, building plausible scenarios, and stress-testing choices before committing resources. Alongside that, we draw on futures design to communicate implications, prototype possibilities, and help teams and stakeholders make sense of complex change, without confusing concept sketches for certainty.

Artificial intelligence now accelerates this work. Used thoughtfully, AI scales horizon scanning, helps structure weak signals, drafts scenario narratives, and generates quick visualizations and simulations. But acceleration is not the same as understanding. Human judgment—values, ethics, context—remains the governor of quality. Throughout these pages you will see workflows where AI does the heavy lifting on volume and speed while humans do the work that only humans can: asking better questions, weighing trade-offs, and building legitimacy with the people affected by our decisions.

This book is pragmatic. It favors clear steps over jargon and shows how to move from abstract frameworks to concrete choices. You will learn to set a sharp intent, run 360° environmental scans (e.g., looking across social, technological, economic, environmental, political, and values factors), cluster signals into patterns, and turn those patterns into scenarios that reveal risks and opportunities. You will practice implications mapping and lightweight futures wheels to surface second- and third-order effects

SEEING FURTHER, ACTING SOONER

that day-to-day planning misses. You will also see where futures design artifacts such as storyboards, mock interfaces, and service maps make strategy more communicable and testable.

Just as important is what this book does not do: it doesn't promise prediction, and it doesn't treat visual artifacts as outcomes in themselves. Where design appears, it is in service of foresight, clarifying options, exposing assumptions, and enabling participation. Where AI appears, it is situated inside guardrails: cite sources, distinguish facts from synthesis, check for bias, and keep a human in the loop.

PART I

The Foundations of Design and Futures Thinking

CHAPTER 1

The History of Design and Contemporary Design Thinking

1.1 Evolution of Design Principles and Practices

The prehistoric landscape of the Lower Paleolithic era gave rise to humans faced with a harsh and unforgiving environment. The need for survival amid predators, changing climates, and the search for food was the mother of invention, akin to the iconic scene from the famous movie *2001: A Space Odyssey*. The need to process food and materials faster and more efficiently was indicative of the need to find ways aiming at the reduction of tasks that are time consuming and physically taxing with bare hands. The men and women of the era noticed that certain types of rocks could fracture to create more durable and sharper edges, experimenting with different stones using striking techniques, from trial and error, discovering the unfolding of their capabilities and their tools, eventually crafting their best effective device: the hand axe.

CHAPTER 1   THE HISTORY OF DESIGN AND CONTEMPORARY DESIGN THINKING

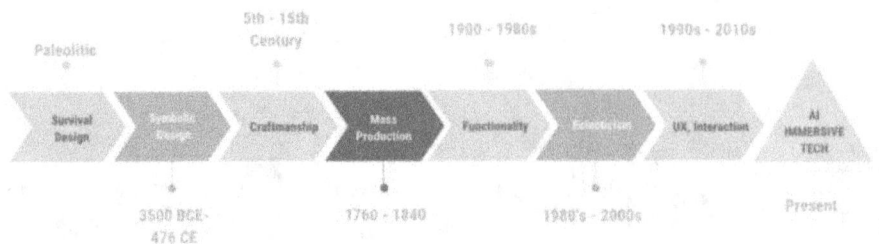

Figure 1-1. History Timeline of Design Thinking

As depicted in Figure 1-1 evolving with their tools, the artisans of the ancient Egypt, Greek, and Roman empires become known for their intricate hieroglyphics and monumental architecture and artifacts, using their knowledge of materials for each best use, shaping the expectations of society, and transforming the role of beauty and art in culture, where we can see how design was influenced by symbolism, functionality, and aesthetics.

A shift occurs in the evolution of design during the Middle Ages and the Renaissance period where a higher emphasis is placed on craftsmanship, stemming from a religious period where some classical principles take center like perspective, depth, proportion, symmetry, balance, and harmony as well as the use of particular motifs, giving birth to key developments in art, architecture, and engineering.

New expressions of art continue to emerge, and toward the mid-18th century, indications of the industrial revolution begin to evolve. With the advent of mass production, new materials and new mechanization methods shaped the creation of products indicating an early separation of design from craftsmanship and individuality—a trend that, toward the 20th century combined with rapid technological advancements and socio-economical changes, sets the stage for the modernism design era, proposing a transition from excess to simplicity and functionality with a special focus on materials, with the introduction of the Bauhaus movement.

With more access to options coming from mass production, a more culturally open society, the postmodern era brings a period characterized by a diversity of styles and a combination of previous art techniques and design methods, reflected in the eclectic works of the time that brought an augmented effect thanks to the digitization of art of design as the first indications of a new major shift in design thinking.

The postmodern era brings us closer to the emergence of more concerned considerations in the field of user experience (UX), interaction design, and human–computer interaction (HCI), highlighting the massive shift from physical to digital products and services, with a particular emphasis on design practices, social responsibility and sustainability, eco-friendly materials, standards and practices, as well as a new focus on ethical frameworks and principles in design.

Contemporary Design, what is coming from last year's shift in technology, is all about AI and immersive tech. At the time of writing this book, for example, we see signs of remote work happening thanks to enhanced interconnected networks and enhanced devices that allow you to interact with digital content like drawings or engineering specs without having to be in person in the same room with the rest of the project team members. We also see more and more immersive tools and now potential increased efficiency with the upcoming introduction of AI agents for our daily work, where we might be brainstorming with a digital assistant to complete a report, get data analysis insights, or even to assemble a presentation.

1.2 Key Movements and Figures in Design History

Throughout history, design has evolved alongside social and technological change. Understanding these shifts provides context for practice today and clarifies which tools and values to carry forward. Historical perspectives

CHAPTER 1   THE HISTORY OF DESIGN AND CONTEMPORARY DESIGN THINKING

reveal recurring patterns and principles that can be re-applied to new challenges, informing both strategic foresight and design decisions. Below is a concise, sourced recap of major movements and figures.

1.2.1 The Arts and Crafts Movement

Led by figures like John Ruskin and William Morris, Arts and Crafts responded to industrialization by re-centering workmanship, materials, and ethical production, laying groundwork for later debates about the unity of art, craft, and industry (Pevsner 2005; MacCarthy 2014). This movement signals how production ethics and transparency shape future adoption, regulation, and brand trust.

1.2.2 Art Nouveau

Art Nouveau rejected rigid academic symmetry in favor of organic, flowing forms and new fabrication techniques; key figures include René Lalique, Antoni Gaudí, and Alphonse Mucha (Silverman 1989; Fiell and Fiell 2017). In Art Nouveau, it becomes evident how material and technical innovations as well as aesthetic shifts can reframe consumer desirability across scenarios.

1.2.3 The Bauhaus Movement

Founded by Walter Gropius, the Bahaus integrated art, craft, and technology in a curricular model that profoundly shaped 20th-century design; notable contributors include Wassily Kandinsky, Marcel Breuer, and Paul Klee (Droste 2019; Wingler 1969). The Bahaus movement demonstrates how education and capability building change the feasible set in futures planning.

1.2.4 Modernism

Ironic for the "form follows function" ethos, modernism emphasized clarity, utility, and material honesty; major figures include Ludwig Mies van der Rohe, Le Corbusier, and Frank Lloyd Wright, who pursued harmony among structure, site, and use (Frampton 2020; Heskett 2005). Modernism clarifies constraints and performance criteria that inform probable futures and design trade-offs.

1.2.5 Art Deco

Characterized by bold geometric shapes and new materials, this art movement influences design by bringing back additional ornamentations and sleek designs using unusual materials like decorative glass, terra cotta, stucco, and metals. Representative designers include Tamara de Lempicka, Eileen Gray, and Jean-Michel Frank (Duncan 2009; Fiell and Fiell 2019). This movement reminds us that cultural mood and technology can align to create new preference spikes in markets.

1.2.6 Mid-Century Modern

Use of natural materials like wooden functional pieces are the centerpiece of this movement along with minimal fuss and ornaments as a result of the evolution coming from the post World War II design trajectory. Influential figures include Eero Saarinen, George Nelson, and Charles and Ray Eames (Framton 2020; Fiell and Fiell 2019). Mid-Century Modern highlights how demographics and housing policy reshape probable use contexts and product service ecosystems.

1.2.7 Postmodernism

The postmodern movement brings an air of contradiction and complexity, which is why it is known as one of the most debated philosophical movements in design history. It clearly made a departure from the previous movements that highlighted simplicity and function, bringing more chaos and unpredictability to its works provoking thoughts about futuristic apocalypses. This type of design challenged universal rules, introducing plurality, irony, and contextual references. Key voices include Robert Venturi, Michael Graves, and Ettore Sottsass (Jencks 2011, Adamson, Pavitt, and Wood 2011). Postmodernism underscores the need to explore multiple narratives and competing meanings across possible futures.

1.2.8 Sustainable and Contemporary Design

Contemporary practice increasingly integrates sustainability, systems thinking, and social innovation, linking design decisions to lifecycle impacts, ethics, and governance. Aligned with the technological and highly consumerist societies, the environmental impact becomes a fundamental factor that enters the design considerations based on legal regulations and organizational values. Focused on resource-efficient practices and environmentally friendly works and creations, this approach to design aims to integrate socially responsible frameworks to design as well as an emphasis on technology to minimize waste and balance the use of natural resources. Influential contributors include Victor Papanek on socially responsible design, Ezio Manzini on collaborative/social innovation, Richard Buchanan on design thinking's scope, and William McDonough (with Michael Braungart) on circularity (Papanek 1985; Buchanan 1992; Manzini 2015; McDonough and Braungart 2002). Contemporary design makes preferable futures explicit connecting targets such as circulatory or equity to backcasting and governance.

Together, these movements expand the evidence base we consider in foresight, linking historical patterns to future implications for products, services, and systems.

1.3 Contemporary Design Thinking and Futures Thinking

While the roots of design thinking stretch back decades, the current mainstream embrace of this approach is driving exciting new applications and evolutions. At its core, contemporary design thinking upholds the human-centered principles of the past, but adapts them to today's rapidly changing technological landscape and increasingly complex societal challenges.

This evolution naturally connects with futures thinking—a discipline focused on exploring potential futures and their implications on society. While traditional design thinking helps us solve present-day problems, futures thinking encourages us to consider long-term impacts, helping designers to anticipate and shape future possibilities.

Futures thinking involves systematically scanning for signals of change, analyzing trends, and imagining various scenarios that could unfold in the future. For example, a designer might use futures thinking to explore how AI could transform urban transportation, considering scenarios where AI optimizes traffic flow for sustainability versus scenarios where it exacerbates social inequalities. By integrating futures thinking into the design process, we can proactively address potential challenges and opportunities, rather than reacting to them as they arise.

AI and sustainability are central to this approach. As AI technologies continue to evolve, futures thinking provides a framework for exploring their potential impacts on sustainability. For instance, through futures thinking, we might envision how AI could be used to enhance energy efficiency in smart cities, reducing carbon footprints and promoting environmental sustainability. Conversely, we might also explore scenarios where AI-driven automation leads to increased resource consumption and waste. These explorations help guide design decisions that align with long-term sustainability goals.

This connection between contemporary design thinking and futures thinking is crucial for navigating today's complex and fast-paced world. By adopting a futures-oriented mindset, designers can better understand how their current decisions might influence future outcomes, allowing them to create solutions that are not only innovative but also resilient and adaptable to change. This forward-looking design process serves several critical purposes. It sparks deeper dialogues about ethical considerations and guides innovation toward more sustainable and equitable outcomes as it engages diverse stakeholders in co-creating visions and solutions for charting preferred paths forward.

For designers, adopting a futures mindset opens up vast new creative playgrounds for exploring speculative "what if" scenarios. But futures thinking requires an expanded design thinking toolkit—from environmental scanning and trend analysis to systems thinking and storytelling capabilities. Designers must become just as versed in the driving forces shaping the future as they are in human needs and behaviors.

Contemporary design has already started embracing futures design, but there is a need for more designers to develop this vital skill. Those who cultivate futures literacy will be invaluable partners for organizations and civic leaders aiming to get ahead of the curve on technological disruption. They will be uniquely positioned to drive ethical, human-centered innovation in an exponentially changing world.

1.4 Summary

This chapter traced the evolution of design from its origins in prehistoric tool-making to the sophisticated practices of ancient civilizations and the transformative shifts brought about by the Industrial Revolution and Modernism. As we move into the contemporary era, design has embraced digital technologies, leading to the integration of AI and immersive tools.

We also explored how contemporary design thinking has expanded to include futures thinking. This approach equips designers to anticipate and shape future scenarios, ensuring that innovations are guided by ethical and sustainable principles, particularly in the rapidly evolving landscape of AI.

CHAPTER 2

The History of Futures Thinking

2.1 Origins and Evolution of Futures Thinking

While futures design is a relatively new practice, the act of speculating about the future has ancient roots. Humans have always tried to anticipate what lies ahead, whether by examining natural patterns, interpreting prophecies and omens, or simply wondering "what if?"

Some of the earliest documented examples of futures thinking come from antiquity. The ancient Sumerian epic of Gilgamesh, dating back to around 2500 BCE, explores themes of immortality and humanity's future. Around 600 BCE, the ancient Greek historian Herodotus was one of the first to systematically analyze the causes of events to extrapolate future possibilities.

As the scientific revolution took hold in the 17th century, rigorous methods for probabilistic forecasting emerged. Mathematicians like Girolamo Cardano and Blaise Pascal developed foundations for probability theory and decision theory. This allowed futures thinking to evolve from prophesying to quantitative modeling and risk analysis.

CHAPTER 2   THE HISTORY OF FUTURES THINKING

The 20th century saw further professionalization of the futures field, termed "futures studies" (often called "futurology" at the time). Figures like H.G. Wells, Alvin Toffler, and The World Future Society popularized studying the future through observation of social trends and technological developments. Corporations and governments established strategic foresight units to apply scenarios and systems modeling.

Parallel to these developments, the design field increasingly recognized the need to create user-centered products and services aligned with future contexts and behaviors. Writings by pioneers like Victor Papanek advocated for designers to become "futurists" by better anticipating societal needs and challenges. This set the stage for design futures to emerge as a formal area of practice.

In the 21st century, accelerating social, technological, and environmental change has structured futures work indispensable. Alongside human-centered practice, contemporary foresight also engages more than human perspectives and infrastructures. This broadens what counts as evidence and value in scenarios, and it reframes "preferable" futures in terms of ecological viability, justice, and care, not only human utility (Escobar 2018; Constanza Chock 2020; Pig de la Bellacasa 2017).

Contemporary practitioners draw upon a rich multidisciplinary toolkit including ethnography, systems thinking, science fiction prototyping, and experiential scenario planning. What unifies these approaches is using speculation to make apparent the relationship between present actions and future impacts.

As the pace of change intensifies, mastering futures literacy—the ability to think strategically about potential futures—will be crucial for navigating uncertainties and steering toward more preferred outcomes for humanity.

2.2 Major Theorists and Approaches in Futures Studies

The field of futures studies has been shaped by many influential thinkers who have pioneered different philosophies, methods, and schools of thought. As represented in Figure 2-1, these are some of the major theorists and approaches that are worth to know about due to their contributions to futures studies:

The Prospectivists: Thinkers like Gaston Berger, Bertrand de Jouvenel, and Michel Godet developed la prospective—a holistic approach to anticipating alternative future scenarios. It emphasizes exploring relationships between events, trends, and actors to uncover potential future paths.

The Probabilists: Theorists like Theodore Gordon, Olaf Helmer, and other RAND Corporation researchers advanced probabilistic forecasting models using cross-impact analysis, trend impact analysis, and other quantitative techniques to assign probabilities to future events.

The Intuitionists: Figures like Joseph Coates, Robert Jungk, and Alvin Toffler emphasized using intuitive thinking, such as science fiction storytelling and experiential scenario planning, to expand the boundaries of how we envision the future. The intuitionists believe pure data analysis has limits in anticipating paradigm shifts.

CHAPTER 2 THE HISTORY OF FUTURES THINKING

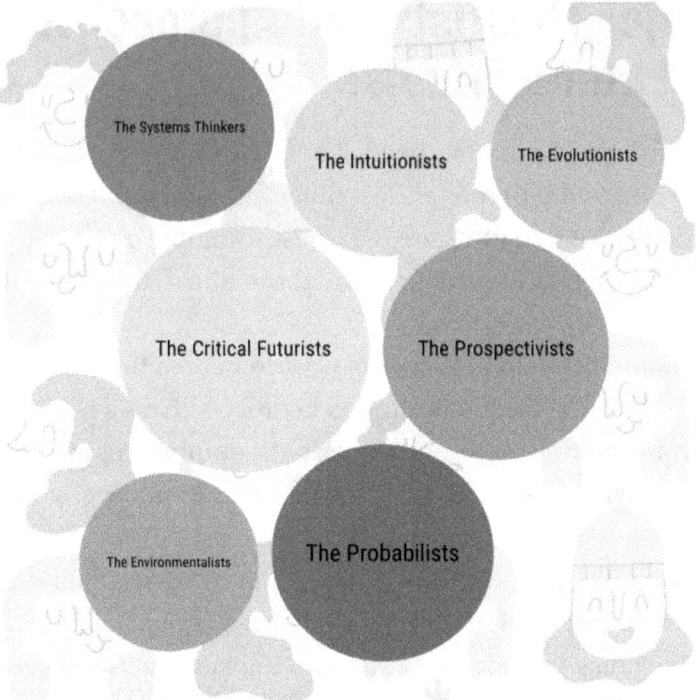

Figure 2-1. *Major Theorists in Futures Studies*

The Evolutionists: Thinkers in this camp, such as Fred Polak, see images of the future as the core driving force behind cultural evolution and social change. They study how collective "Images of the Future" shape human values, behaviors, and actions over time.

The Systems Thinkers: Approaches from systems theory, operations research, and cybernetics have strongly influenced futures studies. Theorists like Russell Ackoff, C. West Churchman, and John N. Warfield pioneered systems thinking for better understanding interconnections shaping the long-term future.

The Environmentalists: With roots in The Limits to Growth from the Club of Rome, theorists like Dennis Meadows, Donella Meadows, and Jorgen Randers advanced modeling approaches to study the future impacts of global environmental and socioeconomic trends.

CHAPTER 2   THE HISTORY OF FUTURES THINKING

The Critical Futurists: Emerging in the 1980s–1990s, theorists like Sohail Inayatullah, Richard Slaughter, and Jennifer Gidley called for decolonizing the future by centering diverse cultural viewpoints. This camp aims to surface alternative futures marginalized by dominant power structures.

Contemporary futures studies practice synthesizes aspects of these different schools of thought. An ongoing tension exists between quantitative and qualitative methods, between predictive modeling and exploratory storytelling. But most futurists agree on the need for multidisciplinary approaches to map our complex, uncertain futures.

2.3 Integration of Futures Thinking in Modern Disciplines

While futures studies emerged as a transdisciplinary field in its own right, elements of futures thinking have increasingly become embedded within other academic disciplines and professional domains. As the pace of technological and social change accelerates, the ability to anticipate and strategize for future possibilities has become invaluable across multiple sectors, and in today's world, we see in North America and Europe some of the most applicable cases in these fields.

Design and Innovation: As explored earlier, design futures and speculative design represent the integration of futures thinking into the design process. Methods like scenario planning, science fiction prototyping, and experiential futures are now taught in design curricula and practiced within strategic innovation teams. With the advent of AI and immersive technologies, these methods have gained new dimensions. AI-driven tools can now simulate complex scenarios with unprecedented accuracy, allowing designers to prototype not just physical objects but entire systems of interaction. Immersive technologies, such as virtual and augmented reality, enable stakeholders to experience potential futures in a visceral way, fostering deeper understanding and engagement. This integration is crucial as it allows designers to explore the ethical implications and sustainability of their innovations in ways that were previously impossible.

Business and Management: Corporate foresight programs, using approaches like environmental scanning, trend analysis, and alternative futures modeling, allow companies to better identify emerging risks and opportunities. Futures thinking informs areas from new product development to supply chain resilience planning. Incorporating AI and big data analytics into these processes has enhanced the ability to predict and respond to market shifts, while immersive tools provide new ways to visualize and strategize around complex business scenarios. The necessity of this integration lies in the rapidly changing global market landscape, where businesses must innovate not only to stay competitive but to ensure sustainable growth. The preliminary risks include over-reliance on AI predictions without considering qualitative insights, which could lead to strategic blind spots.

Technology Governance: Rapid advances in fields like AI, genetics, and computing are driving the need for responsible innovation road mapping. Futures thinking allows researchers, ethicists, and policymakers to analyze potential impacts of new technologies and develop ethical governance frameworks. AI can process vast amounts of data to model future scenarios with high precision, helping to forecast the societal impacts of technological advances. However, the integration of AI also introduces risks, such as the perpetuation of biases or the unintentional creation of harmful technologies. Therefore, the use of AI in futures thinking must be paired with strong ethical oversight to ensure that innovation aligns with societal values and sustainability goals. Immersive technologies can facilitate stakeholder engagement by providing interactive platforms for discussion and decision-making, making complex future scenarios more accessible and understandable.

Sustainable Development: Climate change, biodiversity loss, and resource depletion underscore the urgent need to envision sustainable future systems. Futures methods like The Manoa Scenario process are applied in developing long-term environmental policies and resilience planning. AI can significantly enhance these processes by providing predictive analytics for climate models and resource management, while immersive technologies can help visualize the impacts of environmental decisions on future generations. The integration of these technologies in sustainable development is essential to create more informed, effective, and transparent policies. The risks include the potential for technology to oversimplify complex environmental issues or to be used to justify unsustainable practices under the guise of innovation.

Urban Planning: The future of cities and communities is being shaped by anticipatory and participatory processes that envision alternative scenarios of infrastructure, mobility, housing, and public services. Initiatives like Future Cities Catapult bring futures thinking into urban development. AI and immersive technologies are transforming urban planning by enabling the simulation of urban growth patterns, transportation networks, and environmental impacts in real-time. These tools are necessary to create adaptive, resilient cities that can respond to the challenges of the future, such as climate change, population growth, and resource scarcity. However, reliance on these technologies must be balanced with community input and ethical considerations to avoid creating cities that serve only the needs of the privileged or exacerbate existing inequalities.

Education and Learning: Fostering futures literacy—the ability to envision and prepare for possibilities—has become a focus in modern pedagogies. Approaches like Learning from the Future engage students in using foresight to confront grand challenges. AI-powered educational tools can personalize learning experiences and provide students with real-time feedback on complex future scenarios, while immersive technologies offer interactive, experiential learning environments. The integration of these technologies is crucial for preparing future generations to navigate and shape the world they will inherit. However, the risks include the potential for technology to overshadow critical thinking skills or to exacerbate educational inequalities.

Across these disciplines, two common threads unite the use of futures thinking. The first is the imperative to move beyond insular, short-termist thinking to confront the long-term, large-scale, and interconnected forces shaping our collective future. The second is the belief that the future is not predetermined, but can be positively influenced by intentional vision, strategy, and effort in the present.

2.4 Summary

This chapter explored the key theorists who have shaped the field of futures studies, from the Prospectivists and Probabilists to the Critical Futurists. Each group has contributed unique methods and philosophies, from probabilistic forecasting to intuitive scenario planning, all aimed at better understanding and influencing future possibilities.

We also examined how futures thinking has been integrated into modern disciplines such as design, business, technology governance, sustainable development, urban planning, and education. The integration of AI and immersive technologies into these areas enhances our ability to anticipate and shape future scenarios, driving innovation while addressing ethical and sustainability challenges. However, the use of these tools must be approached with caution to ensure they contribute positively to a sustainable and equitable future.

PART II

Exploring Futures Design

CHAPTER 3

Core Concepts in Futures Thinking

3.1 Defining Futures Thinking: Beyond Traditional Practices

Often, while having discussions about the focus of futures thinking with other designers or technologists, I have identified myself as being misunderstood as if I was attempting to predict the future, which led me to the realization that the reason for this confusion is the name in itself: Futures. In reality, futures thinking is not about predicting a single future, but rather about exploring multiple potential futures that can inform solid decision-making in the present. This approach differs from predicting a single future and therefore anticipating only one outcome. In today's world we can see many futurists like Jim Dator, who recognize that the future is not predetermined, but rather shaped by interacting forces, often complex and subtle as well as the decisions that we make in the present.

Another common misconception is that futures thinking is the same but with a different name than traditional practices like forecasting, strategic planning, and risk management, which mostly rely on linear extrapolations of past trends attempting to deliver better results in short term and relatively stable environments. Futures thinking, however,

CHAPTER 3   CORE CONCEPTS IN FUTURES THINKING

operates on a longer-term horizon, emphasizing the need to navigate uncertainty and complexity by considering a broader range of possibilities. It sets the stage not necessarily in the past—although it can inform the decision-making—but its tooling is framed around the study of current signals and indicators about potential unfolding scenarios toward long-term planning.

Let's take a look at some ways that futures thinking expands the limits of the traditional practices in the studies of the future to inform decision making strategies:

Systems Thinking: As a holistic practice, futures studies have a systems-oriented approach. This means that rather than isolating individual trends, futures thinking considers the interconnections of driving forces, examining how they influence each other over time. As with natural processes, futures thinking considers the interconnections of driven forces in relationship with trends and the resulting influences over a timeframe. Mapping tools here serve as a supportive mechanism to help visualize these complex relationships and the feedback loops required for detailed analysis of data. For example, in urban planning, systems thinking can help identify how changes in transportation technology might impact housing, employment, and environmental sustainability, providing a more comprehensive basis for decision-making.

Participatory Exploration: Futures thinking embraces social responsibility and our place in the world as individuals and as a thriving society. To handle the complexity and inherent ambiguity of long-term change, it recognizes that diverse perspectives are crucial. It recognizes that everyone has a protagonist role in the future, leveraging participatory methods that build collective futures agency and literacy by engaging diverse stakeholders in co-creating visceral scenarios. Participatory exploration ensures that the futures we envision are inclusive and representative of various societal needs, making the outcomes more resilient and adaptable. This approach is particularly powerful when combined with immersive technologies like VR, which allow participants to experience and interact with possible futures, deepening their engagement and understanding.

Futures Typology (PPPP): In strategic foresight, we explore a portfolio of futures—the possible (everything we can imagine), plausible (consistent with what we currently know), probable (what is likely if trends continue), and preferable (what we would choose to pursue given our values and constraints).

Preferred Futures: After mapping possible, plausible, and probable contexts, teams can articulate preferable trajectories, outcomes worth pursuing given ethical, social, ecological, and economic criteria. Preferences are plural and contested; what is preferable for one group or species may not be for another. We therefore treat "preferable" as co-created, and evidence linked: criteria are made explicit, stakeholder (including more than human considerations) are engaged, and we use backcasting to work from those criteria to today's choices. In practice, AI can help surface signals and evaluate trade-offs, but human judgement sets the criteria and accountability.

3.2 The Emergence of Scenario Planning and Speculative Design

As futures thinking evolved to embrace systems thinking, participatory exploration, and the pursuit of preferred futures, new methodologies and practices emerged to operationalize these principles. Two influential approaches that gained traction are scenario planning and speculative design.

CHAPTER 3   CORE CONCEPTS IN FUTURES THINKING

3.2.1 Scenario Planning

Scenario planning is a structured way to imagine and articulate multiple plausible futures. It emerged in the 1960s pioneered by futurists like Herman Kahn at the RAND Corporation, who recognized the inherent uncertainty and complexity of the future. Rather than pursuing a single predicted future, scenario planning explores various "what if" scenarios based on different potential trajectories of key driving forces.

The scenario planning process typically involves

1. Identifying focal issues and decision areas of strategic importance

2. Mapping out the key forces, trends, and uncertainties that could shape the future landscape

3. Constructing a set of divergent yet plausible scenarios around potential interactions of these forces

4. Analyzing the implications and consequences of each scenario

5. Using these insights to inform strategic planning and prepare robust strategies across scenarios

By considering multiple scenarios, organizations can better anticipate disruptions, identify potential risks and opportunities, and develop strategies resilient to various future contexts. Famous examples include Shell's pioneering use of scenario planning to navigate oil crises and Monsanto exploring agriculture scenarios influenced by factors like climate change.

3.2.2 Speculative Design

While scenario planning focuses on exploring multiple potential futures narratively, speculative design takes these ideas into the conceptual and artifact space. Emerging in the 1990s, speculative design uses provocative prototypes, products, and services to spark conversations about different futures we could face.

Speculative designers like Anthony Dunne, Fiona Raby, and others create intentionally fictional and future-oriented designs that embody different values, social premises, and technological possibilities. These 'design fictions' make tangible the potential impacts and implications of different futures in visceral, experiential ways.

For instance, Dunne & Raby's 'Foragers' envisions a post-apocalyptic future where society has reverted to hunting and gathering lifestyles, sparking reflection on sustainability and technology dependence. The 'Rayfish Footwear' by Auger-Loizeau considers a Bio-Cyber future of hybrids between technology and biology. Such examples use design as a discursive space to debate alternative scenarios before they unfold.

As a recent example, Figure 3-1 shows a contemporary speculative design work used to provoke discussion about near-term societal and ecological futures.

CHAPTER 3   CORE CONCEPTS IN FUTURES THINKING

Figure 3-1. Example of Contemporary Speculative Design; Alexandra Daisy Ginsberg, **Pollinator Pathmaker** *(2021–ongoing). Living artwork designed for pollinators, used to explore more-than-human preferable futures.* **Image credit:** *Alexandra Daisy Ginsberg/ Pollinator Pathmaker. Courtesy of the artist and venue*

By crafting artifacts from different potential futures, speculative design expands the possibilities we can comprehend and provokes re-imagining our relationships with emerging technologies, environmental issues, and societal trajectories. It democratizes futures thinking by making it tangible and engaging diverse audiences.

Both scenario planning and speculative design emerged as complementary ways to apply futures thinking principles. They empower organizations and society to explore multiple potential futures and make deliberate choices to shape preferred trajectories proactively.

3.3 The Evolving Role of Strategic Foresight in Design

As futures thinking principles and methodologies like scenario planning and speculative design gained traction, their applications began extending into various domains, including the field of design. This integration gave rise to the practice of strategic foresight in design.

Strategic foresight combines futures thinking approaches with design principles and creative processes. It represents a shift from the traditional design focus on solving existing problems to also exploring potential future contexts and opportunities. By anticipating emerging trends, signals, and disruptions, strategic foresight allows designers to get ahead of the curve.

The role of strategic foresight in the design process involves the framing of future contexts using futures thinking methods like environmental scanning, trend analysis, and scenario development, which designers can use to map out key forces shaping the future landscape relevant to their design domains. This framing surfaces potential future scenarios as contexts to design for. This process leads to the process of envisioning possibilities that follows from rich future contexts established, used by designers to tap into their generative strengths. They apply speculative design approaches to visualize and prototype ideas for how products, services, environments, etc., could evolve to address emerging needs in different scenarios.

CHAPTER 3   CORE CONCEPTS IN FUTURES THINKING

As discussed in the previous chapter, directing toward preferred futures, we realize that strategic foresight isn't just about exploring the future, but actively shaping it toward more preferred trajectories. This leads designers to crystallize visions for desirable futures, then map out the stepping stones and transition pathways to bridge innovation from present to future, which in turn allows for more future-focused design processes, as foresight becomes integral, facilitating design processes themselves to evolve toward a more future-oriented outcome from the outset. Methodologies like transition design explicitly incorporate futures thinking into the full design cycle—from the initial research and sensemaking phase through iterative cycles of prototyping future concepts and solutions.

Across sector, healthcare devices, consumer technology, mobility, public services, and finance, organizations are institutionalizing strategic foresight inside design. Dedicated teams run recurring cycles of scanning ➤ sense making ➤ scenarios ➤ artifacts ➤ backcasting, working with product, policy and Research and Development (R&D). Typical outputs include opportunity maps, assumptions logs, scenario artifacts for stakeholder deliberation, and decision briefs that link near-term choices to long-term trajectories. Professional education and university programs now train designers in these methods, and many organizations explicitly include ethics, environmental impacts, and more human considerations (ecosystems, infrastructure, non-human stakeholders) in their criteria for "preferable" futures. In practice, this raises design's remit from making "things" for users to shaping socio technical systems with affected communities. Some examples in practice include medical devices with scenarios on home diagnostics, following by opportunity maps and backcasting to regulatory milestones, or urban mobility with signals of micro logistics using 2x2 scenario sets, following by service blueprint prototypes for policy review.

3.4 Summary

This chapter explored the evolution of futures thinking in design, focusing on scenario planning and speculative design. Scenario planning helps organizations envision multiple plausible futures to build resilient strategies, while speculative design uses provocative prototypes to engage audiences in imagining alternative futures. The chapter also highlighted the role of strategic foresight in design, where futures thinking is integrated to anticipate and shape future contexts, enabling designers to craft solutions that align with preferred, sustainable trajectories.

CHAPTER 4

Global Perspectives and Cultural Impact

4.1 Diverse Cultural Viewpoints in Futures Thinking

Why this matters:
Perspectives on the future are shaped by place, history, and values. Foresight works best when those lenses are made explicit so what counts as possible, plausible, probable, and preferable is discussed rather than assumed.

How we use culture in practice:
We translate cultural ideas into method choices (who to involve, what counts as evidence/value, how we backcast). We avoid umbrella labels (e.g., "Global South") and name specific traditions or regions. European perspectives are treated as plural strands among many, and we include more than human considerations where relevant.

What follows:
Brief examples such as Aotearoa New Zealand (Māori), Andean regions, Southern Africa (Ubuntu), Japan, and European strands show how different lenses shape scenario evaluation and design criteria. These are illustrative, not exhaustive.

CHAPTER 4   GLOBAL PERSPECTIVES AND CULTURAL IMPACT

Aotearoa New Zealand (Māori):
Futures are often framed as continuity and stewardship of relationships among people, place, and ancestors. Concepts such as tikanga, whakapapa, and kaitiakitanga foreground obligations to more than human worlds and intergenerational care (Mead 2003).
Implication for Practice:
Evaluation criteria emphasize reciprocity, guardianship, and long-term relational impacts.

Andean Regions (Ecuador/Bolivia):
Buen vivir/sumak kawsay assesses futures by collective well-being and the rights of nature, not only growth or control (Gudynas 2011; Escobar 2018).
Implication for Practice:
Backcasting targets center sufficiency, reciprocity, and ecological limits.

Southern Africa (Ubuntu):
Through Ubuntu ("I am because we are"), preferable futures stress mutuality, dignity, and repair across communities (Metz 2011).
Implication for Practice:
Scenarios and metrics shift from individual utility to relational ethics and shared flourishing.

Japan:
Ideas such as "ma" (the meaningful interval) and kata/katachi (form/pattern) highlight rhythm, process, and refinement.
Implication for Practice:
Teams privilege iterative alignment with context over a single linear "end state," affecting staging and governance.

European Strands:
Some European traditions emphasize rational inquiry, scientific method, and progress narratives, one trend among many (alongside Romantic, environmental, and social democratic currents).

Implication for Practice:
Treat European perspectives as plural and situated, not a single "blank state" view.

The takeaway from these illustrative examples is that when teams say out loud what cultural assumptions they are using, their foresight work becomes more trustworthy and easier to learn from. It stops everyone from aiming at one assumed "normal" future and, instead, helps the group agree on shared criteria (what "good" looks like) and be clear about the trade-offs they are choosing.

4.2 Designing for a Multifaceted World

Through my journey exploring futures thinking across different cultures, I've come to realize just how multifaceted and pluralistic the world we're designing for truly is. The futures we envision cannot be singularly defined, but rather woven together from the diverse narratives, values, and lived experiences of people everywhere.

One perspective that really struck me was studying the remarkable traditional African architecture and settlement design philosophies. This concrete example comes from Sudano-Sahelian earthen urbanism in Mali/Burkina Faso/Niger: thick adobe walls, deep overhangs, and small apertures provide passive cooling, and community plastering rituals (e.g., in Djenné) maintain buildings collectively, an approach that links climate performance with social cohesion (Elleh 1997). Along the Swahili Coast in Kenya/Tanzania, historical coral rag houses in Lamu and Zanzibar use shaded courtyards, cross-ventilation, and street-facing baraza benches to moderate heat and support communal life (Elleh 1997). In northern Cameroon, Musgum teleuk domed earthen dwellings, and in South Africa, Zulu beehive huts show for form and joinery, and local materials are tuned to wind, rain, and mobility needs (Prussin 1995). At the settlement scale, many compounds and towns exhibit self-similar fractal organization, from

35

courtyard clusters to district patterns, supporting way finding, growth, and governance (Eglash 1999). Rather than externally imposed, these evolved through deep community participation and an innate attunement to the local environment. The resulting circular dwellings and fractal-like layouts were optimized for natural airflow, sustainable materials, and cultivating interdependence. Solutions that modern urban planners and architects could surely learn from, had we not been so fixated on Western modernist ideals for so long.

Experiences like this revealed how my own design training was still baked in norms that failed to account for the true breadth of human ingenuity and possibility. It made me reflect on how desperately we need to decentralize design beyond Euro-American industrial centers and include perspectives from under represented regions and indigenous communities, alongside other voices historically excluded from decision-making.

But it's not just about representation. We need expanded design methodologies that interweave analytical and intuitive ways of knowing. Methods that bridge human-centered thinking with nature-centered, spirit-centered, and more-than-human perspectives (ecosystems, non-human species, and infrastructures) remind us of our interdependence with all life.

The more I explore, the more I've come to appreciate the true complexity and multifaceted nature of the world we operate within as designers. There is no singular, universal future to design toward, but rather an infinite plurality of "pluriversal futures" as theorists like Arturo Escobar conceptualize. Worlds shaped by diverse cultural narratives and ways of experiencing reality that have been marginalized for far too long by the homogenizing forces of modernity.

This demands a reframing of the designer's identity and role. We must move beyond the outdated perspective of all-knowing designers creating solutions from an external vantage point. Instead, we need to become co-creators—practicing humility and cultivating the wisdom found in other ways of knowing, being, and relating to this multifaceted Planet we all inhabit. It's the only way we can hope to engender futures that are truly just, sustainable, and life-sustaining for all.

4.3 Case Studies: Global Futures Design

This case study, drawn from "Strategic Foresight Studio: A First-Hand Account of an Experiential Futures Course" by Jake Dunagan and colleagues, from their collective work "Design Futures", exemplifies how futures thinking can be effectively taught and applied in an educational setting, providing valuable insights for designers and innovators.

The Strategic Foresight Studio offers a rich example of how diverse futures methodologies can be integrated into a comprehensive learning experience. While the course is educational, its methodologies and lessons can be applied in any organizational context, regardless of size or type. By introducing this case study, we get to see how every designer or team member can become a better-prepared leader by understanding and contributing this point of view to their team. This study is a good guideline for companies and individuals seeking to foster innovation and resilience.

Strategic Foresight Studio
Course Structure and Objectives
The Strategic Foresight Studio, as described by Jake Dunagan and colleagues, is designed to equip students with the skills and knowledge necessary to navigate and shape the future. The course is structured around a series of workshops, lectures, and hands-on projects that immerse students in the principles and practices of futures thinking.

The primary objectives of the course are

- To introduce students to various futures methodologies, including scenario planning, design fiction, and experiential futures.

- To develop students' ability to think critically and creatively about the future.

- To foster collaborative skills through group projects and discussions.

- To provide practical experience in applying futures thinking to real-world challenges.

Methodologies Used

The course incorporates a range of futures methodologies, each chosen for its ability to illuminate different aspects of futures thinking and strategic foresight.

1. Scenario Planning:

   - Scenario planning is used to help students imagine and analyze multiple possible futures. This methodology involves creating detailed narratives about different future scenarios based on varying assumptions and trends.

   - Students learn to identify key drivers of change, explore their potential impacts, and develop strategies to address potential challenges and opportunities.

2. Design Fiction:

   - Design fiction involves creating speculative artifacts from the future, such as prototypes, stories, and experiences. This approach helps students visualize and engage with potential futures in a tangible way.

   - Through design fiction, students explore the implications of emerging technologies and societal changes, fostering a deeper understanding of how these elements might shape the future.

3. Experiential Futures:
   - Experiential futures focus on creating immersive experiences that allow participants to "live" in a possible future. This methodology emphasizes the sensory and emotional aspects of futures thinking.
   - Students design and implement experiential scenarios, such as installations or performances, to bring future possibilities to life and provoke reflection and discussion.

Student Projects

Throughout the course, students work on a variety of projects that apply the methodologies they have learned. These projects serve as practical exercises in futures thinking and strategic foresight, enabling students to experiment with different approaches and develop their skills.

Example Projects:

1. **Urban Futures:**
   - One group of students created a scenario for a future city where AI-driven infrastructure optimizes energy use and transportation. They developed a series of artifacts, including maps, models, and interactive displays, to illustrate how this future city would function and the challenges it might face.

2. **Healthcare Innovation:**
   - Another project focused on the future of healthcare, exploring the potential of AI and biotechnology to revolutionize medical treatment. Students designed a speculative clinic of the future, complete with patient experiences and futuristic medical devices, to demonstrate the possibilities and ethical considerations of such advancements.

3. **Climate Resilience:**
   - A team of students addressed the issue of climate change by developing scenarios for resilient communities. They used scenario planning to outline different pathways for adaptation and mitigation and created an immersive exhibition to engage the public in their findings.

The following criteria and instruments were used to evaluate whether the studio achieved its aims of building transferable capability.

Evaluation:
In this studio, "success" is defined as the capability build and transferred, evidenced by

- **Method Competence:**
  - Students can run a complete cycle (Scanning ➤ Sensemaking ➤ Scenarios ➤ Artifacts ➤ Backcasting) and make their evidence chain traceable.

- **Quality of Outputs:**
  - Scenario sets, artifacts, and decision briefs meet a rubric for clarity, plausibility, inclusivity, and ethics/more-than-human considerations.

- **Stakeholder Uptake:**
  - External reviewers or partner audiences engage with, cite, or request follow ups on the work (e.g., for policy, product, or community conversations).

- **Transfer of Practice:**
  - Learners report applying one or more methods in internships, studios, or early career roles within six months.

How It Is Assessed:

- Pre-post efficacy survey on future methods
- Rubric scoring of final deliverables (see criteria below)
- External feedback notes from critique panels/exhibitions
- Follow-up check-ins on method use after the course

Rubric (Summary Criteria):

1. Traceability: Sources ➤ Insights ➤ Implications are explicit
2. Scenario Quality: Internal coherence; links to signals and drivers
3. Artifact effectiveness: Communicates scenario implications to non-experts
4. Inclusivity and Ethics: Addresses stakeholders (inducing more-than-human) and trade-offs transparently
5. Actionability: Backcasting identifies near-term tests/decisions

Analysis and Discussion

The evaluation focuses on transferable capability, not just classroom outputs. High scores on traceability and actionability correlate with stronger stakeholder uptake in critiques; projects that translate scenarios into clear decision briefs earn more requests from follow-ups. Variability most often appears in inclusivity and ethics (especially making more-than-human criteria explicit), which becomes a teaching priority in earlier weeks.

Limitations and Next Iteration

Evidence comes from a single program and short follow-ups; long-term transfer to practice is under-measured. Next runs add a mid-course traceability clinic, a backcasting mini sprint and a six-month alumni check in template.

Applicability in Organizational Contexts

The same evaluation logic applies in firms: score work on traceability, scenario quality, artifact effectiveness, inclusivity/ethics and actionability, and track update (citations, pilots, policy briefs). Teams that institutionalize this rubric build foresight capability faster.

Educational Implications

Embedding futures methods in design curricula prepares graduates to make evidence-linked decisions under uncertainty. Programs that use experiential cycles, external critiques, and the evaluation rubric above are better positioned to foster innovation and resilience in the age of AI.

4.4 Summary

This chapter showed how cultural lenses shape what counts as possible, plausible, probable, and preferable futures. We illustrated this with region-specific examples and sources—Māori tikanga/kaitiakitanga in Aotearoa New Zealand (Mead 2003), Andean buen vivir (Gudynas 2011; Escobar 2018), Southern African Ubuntu (Metz 2011), Japanese concepts such as *ma* and kata/katachi, and plural European strands—each paired with practice implications for criteria, participation, and backcasting.

Overall, this section emphasized making cultural assumptions explicit when setting evaluation criteria for "preferable" futures, using region-appropriate evidence and stakeholders as well as inducing more-than-human considerations where relevant.

CHAPTER 5

Ethical and Sustainable Design

5.1 Navigating Ethical Complexity in Futures Thinking

As we venture into exploring potential futures and design interventions to shape their trajectories, we inevitably encounter a vast horizon of ethical quandaries. The long-term, wide-ranging impacts we could set in motion by our actions today demand rigorous moral reflection. Yet the multifaceted complexity of futures thinking also reveals the contingency of our ethical frameworks themselves.

In my own journey grappling with this ethical terrain, I've come to appreciate just how slippery and perspective-dependent our understandings of "right" and "wrong" can be. What may be considered ethically sound from one cultural viewpoint could be seen as blatantly unethical from another. Our ethical axioms are shaped by diverse belief systems, histories, and contexts that are all too often artificially universalized by dominant Western moral philosophy.

Take for instance the foundational principle of individual autonomy that guides much of modern bioethical decision-making. Through a liberal democratic lens, defending the rights of the individual to make

CHAPTER 5   ETHICAL AND SUSTAINABLE DESIGN

free choices about their own body and well-being is sacrosanct. But this premise is far from universal. Many other cultural traditions prioritize collective responsibility and social harmony over radical individualism. The perceived ethical violation is enabling choices that could destabilize family, community, or spiritual equilibrium.

Such conflicts extend beyond abstract ideology into materially consequential futures. As biotechnological capabilities rapidly evolve, debates rage around philosophically tangled issues like germline gene editing, life extension therapies, and cognitive enhancements. Reasonable minds can diverge on the ethics of human augmentation based on their culturally grounded moral premises. Yet the pathways we choose to pursue or restrict will fundamentally shape the future physiologies and experiences of our species.

The ethical ambiguities I have grappled with only seem to multiply when we expand our circle of moral consideration beyond the human dimension. Increasingly, environmental philosophers and ethicists are calling us to transcend anthropocentric lenses that place humans as the sole bearers of intrinsic value. They invite us to develop perspectives that account for the profound interdependencies between human and non-human life—prioritizing a holistic ethics of ecological stewardship.

From this vantage point, conventional sustainability paradigms embraced by many futures efforts are themselves ethically fraught. Merely striving for the sustenance and longevity of our own human systems, without regard for their impacts on broader systemic health and resilience, could be viewed as a cynical form of ethical myopia. True sustainability may demand radically rethinking and redesigning human civilization to operate in reciprocal balance with the planet's holistic well-being.

These are just a few of the tangled ethical nuances futures thinkers must navigate. As we peer into the profound uncertainties and high-stakes implications of emerging technological, social, and ecological trajectories, we cannot rely on moral business-as-usual. The practice of futures

CHAPTER 5   ETHICAL AND SUSTAINABLE DESIGN

thinking demands cultivating ethical resilience—nurturing moral humility, cultural pluralism, and evolving ethical sensibilities flexible enough to account for radically different contexts yet to come.

5.2 Sustainability As a Core Principle

As we confront the ethical and existential risks of current trajectories, one principle emerges as a moral and practical imperative for all futures thinking and design endeavors: sustainability. Yet this notion of sustainability must be understood in its fullest, since traditionally, most understanding of sustainability has been focused mostly on environmental preservation.

Through my own journeys exploring different cultural cosmologies, I have come to appreciate sustainability as a holistic ethos of harmonious co-existence, of patterning human activities and systems to be in balance and reciprocity with the flourishing of all life. It is a recognition of our profound embeddedness within the larger ecological systems that sustain us, and a call to realign our designs and technologies to be symbiotic partners in these systems rather than dominators or parasites upon them.

This expanded conceptualization of sustainability challenges many of modernity's dogmas around perpetual growth, resource extraction, and human supremacy over nature. It reveals the inherent unsustainability and ethical breach of economic and social models predicated on the relentless exploitation of our home planet's finitude in service of unchecked consumption and accumulation.

Decades of evidence show that design choices must operate within ecological limits and support justice across communities and species. Findings from climate science (IPCC AR6) and planetary boundaries research indicate that overshoot in climate, biodiversity, and nutrient cycles is already shaping risk and opportunity for design and policy (IPCC, 2023; Rockström et al., 2009; Steffen et al., 2015).

45

CHAPTER 5 ETHICAL AND SUSTAINABLE DESIGN

Rather than treating any single tradition as a universal remedy, we draw actionable principles from multiple knowledge systems, including indigenous stewardship, commons governance, and circular economy, where the evidence is strong. For examples:

Legal Personhood for Ecosystems (Aotearoa New Zealand): The Whanganui River is recognized as a legal person, aligning governance with kaitiakitanga and changing design criteria for infrastructure and land use (Charpleix, 2018).

Biodiversity Outcomes on Indigenous Managed Lands: Global analysis show large overlaps between Indigenous territories and biodiversity priorities, with management practices contributing to habitat integrity (Garnett et al., 2018).

Cultural Fire Management (Australia): Reviving indigenous burning practices reduces catastrophic wildfire risk and supports mosaic habitats, informing design for resilience (Bowman et al., 2020).

Commons Governance for Resources: Design and policy informed by Ostrom's principles improve long-term stewardship of shared systems (Ostrom, 1990).

Circular Design Frameworks: Approaches such as Cradle to Cradle and doughnut economics operationalize material health, reuse loops and social foundations for projects and cities (McDonough and Braungart, 2002; Raworth, 2017).

What This Means for Design:

Instead of declaring a single "core-philosophy," we recommend criteria that projects make explicit and evidence linked:

1. Stay within ecological ceilings (e.g., carbon, biodiversity).

2. Meet social foundations (equity, health, dignity).

3. Honor place-based governance and rights (including indigenous co-management where applicable).

CHAPTER 5   ETHICAL AND SUSTAINABLE DESIGN

4. Plan for more-than-human impacts over multiple time horizons.

5. Make trade-offs transparent using backcasting from these criteria.

This reframing keeps the normative intent (sustainable and just futures) while grounding it in documented practices and assessable criteria for decisions. In practice, this expanded sustainability principle demands a radical evolution in how we approach futures thinking and design itself. It calls us to deeply embrace systems perspectives over linear, box-centric methodologies. To foreground biomimicry and nature's deep intelligence as examples for design rather than attempting to impose industrial, top-down rationalities.

This view compels us to deconstruct the culturally ingrained dualisms between nature and artifact, and instead cultivate sensibilities of nature+artifact hybridity, understanding our creations not as disjunct from ecology but as new folds in its dynamism we must attune to with wisdom. To design as nature does, transforming impermanent material and energy flows into resilient, adaptive shapes that harmonize with the eternal continuum of life's evolutionary unfolding.

Who Steers and by What Levers: Steering is distributed. Design teams and product leaders set roadmaps and requirements; procurement and operations choose materials and suppliers; investors and boards set risk/return and ESG screens; policymakers, city planners, and regulators shape codes, incentives, and land use; standards bodies and certifiers define thresholds; educators and funders pick curricular and research agendas; and communities—including indigenous co-managers—govern place-based rights and uses. Through these concrete levers, roadmaps, procurement specs, investment criteria, codes/permits, standards, land use plans, and curricula/grants, actors decide which trajectories move forward. Using the criteria above (ecological ceilings, social foundations, more-than-human impacts) makes those choices explicit and accountable.

CHAPTER 5 ETHICAL AND SUSTAINABLE DESIGN

5.3 Long-Term Impacts and Responsibilities

One of the most humbling and consequential realizations I have had to confront through my immersion in futures thinking is the tremendous long-term impacts our actions and decisions in the present can catalyze. The shockwaves set in motion by the ideas, technologies, and initiatives we pursue today have the potential to ripple across decades, centuries, even millennia to come.

This stance of radical long-term outcomes forces us to expand our ethical and strategic apertures far beyond the pitiful short-term view that plagues most leadership agendas. It demands that we deeply internalize the generational impacts, effects, and responsibilities of our societal choices and ways in which the frenetic, quarterly outlook economy and politics simply cannot accommodate.

After all, the truly monumental inflections that could irreversibly shape the trajectories of our entire species and planetary eco-systems lie in that long-term perspective. Whether we wisely steer progress in genomics, artificial intelligence, geoengineering, and other world-remaking fields, it could mean the realization of wildly divergent futures—horizons of utopic transcendence or dystopic peril down which all future generations would be irrevocably funneled.

The unsustainability and grave intergenerational injustices encoded in our present systems first became visceral to me through tangible experiences like bearing witness to the long environmental half-lives of our toxic byproducts. Observing lands that industrial activities had rendered radioactive "sacrifice zones" for millennia to come or hearing from frontline communities displaced by rising seas and droughts that their ancestors could scarcely have conceived of inadvertently catalyzing through their own limited survival efforts centuries ago.

CHAPTER 5 ETHICAL AND SUSTAINABLE DESIGN

Encounters like these underscore how many of the consequences of our actions today will be experienced most acutely by inheritors of the futures we bestow: generations who had no say in the individual and collective choices that predetermined their realities, robbing them of their own self-determination. It is a solemn reminder of the sacred intergenerational trust we bear as today's global majority temporally privileged cadre empowered to design the frameworks of tomorrow.

In effect, responsibly engaging in futures thinking and design requires nothing less than stretching our moral philosophies and spheres of consideration across almost unimaginable dimensions as we move forward as a society in a world full of uncertainty.

Whole-System Foresight: To make wise choices, teams should practice whole-system, anticipatory thinking, what Buckminster Fuller called comprehensive anticipatory design science—look at the entire system (resources, constraints, stakeholders), simulate second- to third-order effects across time, and then choose the smallest, highest leverage intervention, his "trim tab" that moves the system toward the agreed criteria (ecological ceilings, social foundations, more-than-human impacts). In practice, this means pairing systems maps+backcasting before committing to design and policy.

This is perhaps the most awesome responsibility and call to humility we face. We must hold in mind not just the immediate impacts of our actions, but their compounding reverberations across decades, centuries—the deep time periods over which human civilization and planetary life itself will unfold along multiple, path-dependent trajectories with benefits and harms unevenly distributed shaped by the choices we make today.

CHAPTER 5   ETHICAL AND SUSTAINABLE DESIGN

5.4 Summary

In this chapter, we delved into the ethical challenges inherent in futures thinking, emphasizing the need for moral reflection as we explored potential futures and design interventions. The text highlighted how ethical frameworks are deeply influenced by cultural contexts, leading to varying interpretations of what is considered right or wrong. These differences become particularly significant when dealing with emerging biotechnologies, human augmentation, and environmental ethics, where culturally grounded moral premises can lead to divergent views on the future. The chapter argued that to responsibly navigate these ethical complexities, futures thinkers must cultivate moral humility, embrace cultural pluralism, and develop adaptable ethical sensibilities.

The chapter reframed sustainability as operating within ecological ceilings and meeting social foundations and then pointed to regenerative design as the next step: practices that actively restore ecosystems, build community capacity and increase long-term vitality rather than only reducing harm. In foresight terms, teams set criteria for restoration and resilience, use backcasting to plan interventions, and make trade-offs explicit, including more-than-human impacts.

PART III

Integrated Futures Thinking Methodologies

CHAPTER 6

Foundational Methodologies in Futures Thinking

The goal of this chapter is to turn futures thinking into decisions and designs. It is a shift from "What will happen" to "What future we are building and what we do next." We define criteria for desirable futures → scan & synthesize (STEEP+V/PESTLE) → build scenarios (Cone, Wheel) → rehearse with artefacts → pick options, triggers, and experiments.

Read the section "Radical Reimagining and Innovation for Desirable Futures" in Chapter 7 for the underlying stance (reframing assumptions), then use Chapter 7 to apply it.

6.1 Introduction to Futures Thinking Concepts and Approaches

At its core, futures thinking is the holistic practice of exploring potential futures in a structured and systematic way. By analyzing key driving forces, trends, and disruptive signals emerging in the present, futures approaches allow us to map out alternative scenarios about how the future

CHAPTER 6   FOUNDATIONAL METHODOLOGIES IN FUTURES THINKING

could plausibly unfold. This thoughtful contemplation of future contexts is not about static prediction, but about cultivating agency to make wise decisions that shape preferable trajectories.

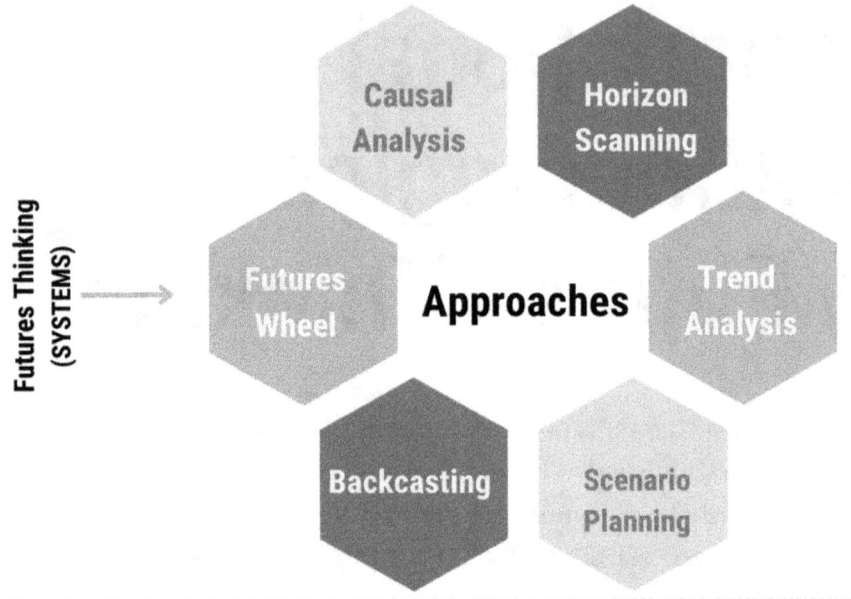

Figure 6-1. Futures Thinking Approaches

Figure 6-1 above illustrates how futures thinking encompasses a diverse toolkit of frameworks, methodologies, and mindsets that can be creatively blended. In Appendix A, you can find a more complete tool box with their respective summaries and when to use each tool, created from the Futures Research Methodology published by the Millennium Project, 3.0 edition. Here, let's begin by studying some of the foundational concepts and approaches:

Horizon Scanning and Environmental Scanning: Structured processes for detecting early warnings and discontinuities that could substantially impact the future landscape across societal, technological, economic, environmental, and political dynamics.

54

CHAPTER 6   FOUNDATIONAL METHODOLOGIES IN FUTURES THINKING

Trend Analysis and Foresight: Analyzing quantitative and qualitative data to map out the potential future impacts and implications of currently observed trends and patterns of change.

Scenario Planning: A core futures technique of developing multiple plausible scenario narratives around how various driving forces may interact and shape alternative future contexts.

Backcasting: Visioning desirable future states, then mapping backwards to strategize the policies, innovations, and transitions required to achieve that trajectory from the present.

Futures Wheel and Impact Analysis: Mapping the cascading potential impacts and knock-on effects that could ripple out from a future event, innovation, or shift across complex systems.

Causal Layered Analysis: A critical futures approach that exposes and excavates the deeper worldview narratives, metaphors, and social constructs that fundamentally shape our conceptions of the future.

In recent years, rapid advances in data analytics, modeling, and artificial intelligence have opened up powerful new possibilities for these foundational futures methodologies. Sophisticated machine learning, pattern recognition, and natural language processing capabilities now augment human futures analysts in tasks like

- Processing and synthesizing vast corpuses of signals data to detect emerging issues
- Visualizing and simulating complex systems dynamics
- Rapidly modeling and running countless "virtual" scenarios
- Generating plausible fictionalized scenario narratives

While powerful, these AI capabilities must be approached with caution and wisdom. They can easily bake in problematic biases, blind spots, and errors from incomplete training data or flawed modeling parameters that then become amplified into futures insights. As such, human intuition, context framing, and ethical scrutiny remain essential counterweights.

CHAPTER 6   FOUNDATIONAL METHODOLOGIES IN FUTURES THINKING

This symbiosis of human and machine intelligence reflects the two complementary streams that have evolved under the broader futures thinking tent:

The first one is Strategic Foresight—which brings futures thinking into organizational strategy, innovation, and planning. Foresight analyzes how issues could unfold, develops wind-tunneling strategic plans across scenarios, and guides proactive responses and interventions.

The second one is Futures Design—an approach focused on speculating and concretizing different potential future worlds, contexts, and artifacts through storytelling, speculative product prototypes, exploratory scenarios, and design fictions. Practitioners aim to make the future tangible and visceral to expand our imagination and provoke discourse.

As we progress through this book, we will explore applications and case studies across both of these streams. We will see how futures design can cultivate aspirational visions, surface blind spots, and spark creativity. And we'll dive into how strategic foresight embeds futures thinking as an ongoing capacity for navigating uncertainty and complexity.

Ultimately, whether through data-driven modeling, experiential sci-fi prototypes, or guided group processes, the promise of futures thinking is to expand our perception of possibility spaces. To transcend the overly deterministic and constrained ideas of "the future" we so often default to. When practiced with rigor yet open-mindedness, futures approaches can rekindle humility to remind us that the future remains malleable and unwritten—profoundly beholden to the choices we make by reflecting on it today.

6.1.1 Overview of Futures Design and Speculative Design Basics

As we journey deeper into futures methodologies, it's important to recognize futures design and its closely linked speculative design research practices. The inputs are systematic: environmental scanning, signal gathering, trend/driver analysis, framing, and scenario development, while the outputs may use artistic media (stories, artifacts, performances) to make evidence-based futures experienceable. These outputs aren't decoration; they are research instruments to surface assumptions, elicit stakeholder feedback, and inform decisions. This approach is well documented in design futures scholarship (Dunne & Raby 2013, Candy & Dunagan 2017; Tharp & Tharp 2019; Auger 2013).

At its heart, futures design is the practice of concretizing potential future scenarios and contexts through tangible stories, models, prototypes, and speculative artifacts. It aims to make potential futures experientially visceral—giving shape and form to emerging possibilities so we can quite literally grasp their contours and implications.

Rather than primarily focused on data-driven analysis and forecasting, futures design prioritizes expansive creativity and unleashing the social power of fiction. Practitioners leverage the digestible familiarity of narratives, speculative products, and design fiction to immerse people in provocative future spaces that transcend the confines of our current paradigms.

This immersion in future storytelling can take many forms. Perhaps constructing a narrative docudrama that illustrates potential trajectories and consequences around an emerging technological shift. Or crafting experiential design fictions like prototyped products and user scenarios that tangibly depict how new socio-technical systems could manifest in the world.

For example, the cult film exemplar of design fiction—Spike Jonze's *Her*—effortlessly primes our awareness to contemplate how we may relate to forthcoming artificial intelligences.

The futures thinking studio Superflux also frequently showcases their explorations in the largest galleries and museums in the world. Recently, they presented their work from their SAFE series "The Seas Are No Longer Dying", illustrated in Figure 6-2 below, on the use of water at the Museum Für Kunst und Gewerbe Hamburg, exploring questions like: Can technologies and infrastructures, systems and temperaments shape a world absent of needless grief and suffering?

Figure 6-2. Superflux "The Seas Are No Longer Dying" Exhibit

Tools like these leverage our innate capacity for imagination. By tangibly instantiating potential futures into experiential artifacts and scenario-worlds we intuit their implications, catalyze discussion, and unlock new openings for stewarding their trajectories.

In the strategic sphere, futures design techniques facilitate envisioning exercises, making possibilities experientially concrete during ideation. They can also foster powerful experience prototypes before investing in fully manifesting innovations or frameworks.

Provocative artifacts and experiential scenarios allow participants to inhabit potential futures and see their ramifications from within. This enables deeper collective understanding and ultimately accelerates innovation pathways by bridging imagination and action.

Whether deployed for catalyzing discourse, enabling transformative visions, or iterative strategy development—futures design empowers us to grasp the malleable nature of forthcoming realities and motivates us to consciously actualize their trajectories in concert with our chosen values and intentions.

6.1.2 Understanding the Future: Definitions, Theories, and Influences

Before we can effectively practice methodologies for exploring the future, it's vital to first ground ourselves in the theoretical foundations and paradigms that shape how we conceptualize the future itself. Our ontological perspectives and embedded assumptions profoundly influence the ways we approach and make sense of futures thinking.

At the most fundamental level, we must confront the basic philosophical question—what exactly is "the future"? Is it a static, predetermined eventuality just waiting to unfold? A branching multiplicity of possibility spaces? Or is it a far more radically indeterminate and ever-receding terrain, remade in each present moment?

Classical theories rooted in scientific determinism and Newtonian physics posited the future as essentially predictable through mastery of initial conditions and causal modeling. In this view, the future represents the culmination of all present forces playing out in an uninterrupted linear trajectory.

However, more contemporary complexity paradigms call this determinism into question. They frame the future as inherently unpredictable and irreducibly complex due to the staggering interdependencies, feedback loops, and emergent properties arising in dynamic systems. The future is not preordained but continually birthed through iterative changes and perturbations.

Influential futures thinkers like Sohail Inayatullah have looked to macro-histories and anthropological inquiry to shed light on how different civilizations and worldviews conceptualize the future. His causal layered analysis reveals assumptions ranging from the cyclical renewal philosophies of ancient Islamic, Indian, and Indigenous thought to the techno-utopian linear visions permeating Western modernism and the more fractured postmodern narratives of today.

These embedded societal metaphors about time, change, agency, and human relationships with nature form powerful root narratives that profoundly shape our images and actions toward the future. Unpacking them surfaces core questions—do we envision the future as predetermined destiny, fertile emergence to steward, or an infinitely malleable space to construct?

A Practice View: Futures can be treated as contingent co-creation arising from intersecting worldviews, projected intentions, and dynamic systemic forces that exceed any individual's control. Rather than a fixed fate, the future is a relational improvisation shaped through imagination and action. This aligns with enactivist perspectives (cognition enacted through embodied coupling with environments) which implies that futures are "brought forth" through situated beliefs, values, choices, and interventions (Varela, Thompson, and Rosch 1991).

From this standpoint, futures methods function as sense-making practices: expanding attention to signals of change; linking evidence to scenarios via analysis and modeling; and using narrative and artifacts to surface assumptions, disrupt limiting frames, and rehearse alternative futures before acting on them.

6.2 Framework for Anticipating Futures

At the heart of most futures thinking practices lies a systematic process for detecting signals of potential disruptions or developments that could shape the dynamics of the future. While the specific steps and techniques used will vary based on the context and needs of an organization, there are some common principles that tend to make for an effective approach to anticipating the future.

The first principle is the importance of establishing a focal point for the effort. By designating a particular achievement, performance indicator, or deliverable as the focal point, it helps to clarify the aim of the anticipation efforts and motivates the entire process. For example, when creating an initiative, the focal point might be presented as: "Producing an analysis report that will inform the strategic direction of the brand Ethos Wear and its sustainable clothing by 2027." Having a clear goal statement like this helps orient the overall effort and facilitates effective prioritization and coordination to meet that goal.

The second key principle is creating a systematic monitoring process to continually scan the environment and detect meaningful signals of change that could affect or reshape the dynamics that matter for achieving the focal aim. Typically, this monitoring process involves

1. Identifying a set of signals, issues, and dynamics that are of interest or relevance to the aim. For example, let's take the following dynamics of interest for other types of efforts: AI ethics governance, a strategy for promoting ethical use of a certain product or service, or brand positioning on ethical AI. By clearly identifying the dynamics of interest, it makes coordination and prioritization easier across the organization.

CHAPTER 6   FOUNDATIONAL METHODOLOGIES IN FUTURES THINKING

2. Assigning clear individuals or teams to monitor the dynamics of interest. This helps create a unified and coordinated system for sharing information across teams and leadership, so that there are monitored owners for each of the dynamics of interest.

3. Establishing monitoring practices and rituals, such as recurring meetings, collaboration spaces, email newsletters, etc., to ensure that updates on the dynamics of interest are consistently shared with monitored owners.

Beyond these basic principles of monitoring to achieve the initiative's goal, it is also important to establish practices and rituals for sharing updates across the organization and with leadership. This opens up the ability of the entire organization to receive periodic updates on the key developments and to be aware of any major developments that need to be escalated to leadership or any significant issues that need to be addressed by leadership or at high-level executive meetings.

By taking these basic steps, the organization creates a systematic and coordinated process for tracking updates on matters of importance to achieve its initiatives and goals, and to ensure automated ownership so that prioritized updates can be shared across all relevant parties. This is a basic requirement for any organization, with the help of their design or innovation teams, to ensure prioritized issues are internally shared with relevant parties so that leadership and stakeholders can be safely kept informed about activities and to facilitate decision-making.

6.2.1 Introduction to the STEEP+V Model and the Futures Cone

6.2.1.1 The STEEP+V Model

Two foundational frames used in scanning are STEEP+V (also written (STEEP+V) and PESTLE/PESTEL. We use STEEP+V in this book (See Figure 6-3 below), because it makes values explicit alongside Social, Technological, Economic, Environmental, and Political factors. In some contexts teams prefer PESTLE/PESTEL (Political, Economic, Social, Technological, Legal, Environmental). The choice is pragmatic; pick the variant that makes your assumptions and evidence most explicit, and state your definitions.

Categories:
Social: Demographics, culture, behaviors.
Technological: R&D, adoption curves, infrastructures.
Economic: Markets, labor, capital flows, incentives.
Environmental: Climate, biodiversity, resources, land/water.
Political: Governance, geopolitics, policy, standards.
+V Values: Norms, ethics, narratives, identity, legitimacy.

(If regulatory detail is central, teams often use PESTLE to foreground legal instead of or alongside +V).

CHAPTER 6   FOUNDATIONAL METHODOLOGIES IN FUTURES THINKING

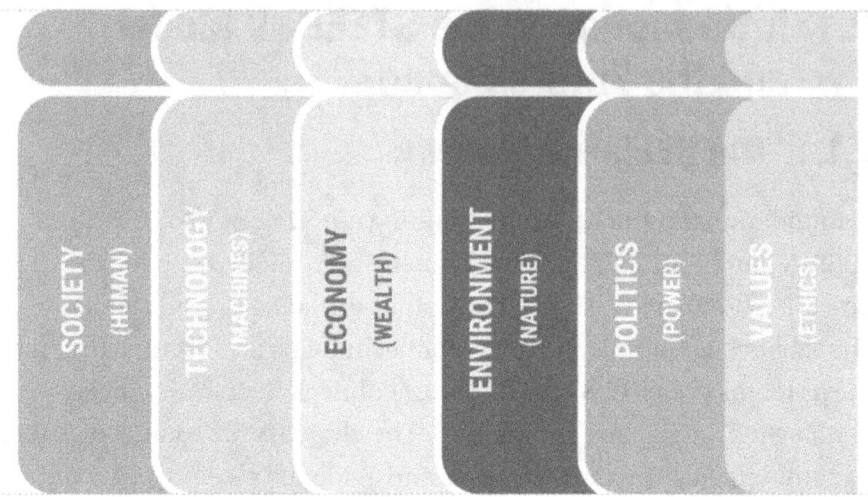

Figure 6-3. *The STEEP+V Model*

The STEEP+V model provides an environmental scanning typology for mapping out the key change drivers across six intersecting domains: Society, Technology, Economics, Ecology, Politics, and Values. By analyzing emerging issues through each of these conceptual lenses, we can develop richer pictures of how different forces may interact to catalyze future shifts and disruptions.

For example, let's consider the nascent development of brain-computer interface (BCI) technologies that will directly link our neural activity to digital systems. Examining this through the STEEP+V framework reveals the following:

Society: BCIs could redefine human–machine relations, augment our cognitive capabilities, and spark debates around identity, embodiment, and autonomy.

Technology: Advances are required in bioelectronics, AI, quantum computing, and brain modeling to make BCIs viable and scalable.

Economics: BCIs open up new industry sectors around neural software/services but could deepen societal divides if access is unequal.

Ecology/Environment: How might our ecological footprints change as our experiences become more virtual? Risks of e-waste from BCI hardware?

Politics: Inevitable battles over privacy, security, and regulatory regimes governing cognitive technology and human enhancement.

Values/Ethics: Fundamental philosophical questions around what constitutes human nature and human identity in a world of mind–machine fusion.

By systematically working through each of these contextual dimensions, the STEEP+V model pushes us to consider the multi-faceted ripples a new technology or trend could generate across different sectors of society. It combats overly siloed thinking to surface potential impacts, challenges, and stakeholder perspectives we may otherwise overlook.

6.2.1.2 The Futures Cone

The Futures Cone (See Figure 6-4 below), is another powerful framework for mapping out alternative future scenarios and possibilities related to any focal issue being analyzed through STEEP+V. The cone metaphor depicts the present as a narrow reality we're embedded in, but which expands into a widening aperture of plausible alternative futures the further out we look.

CHAPTER 6   FOUNDATIONAL METHODOLOGIES IN FUTURES THINKING

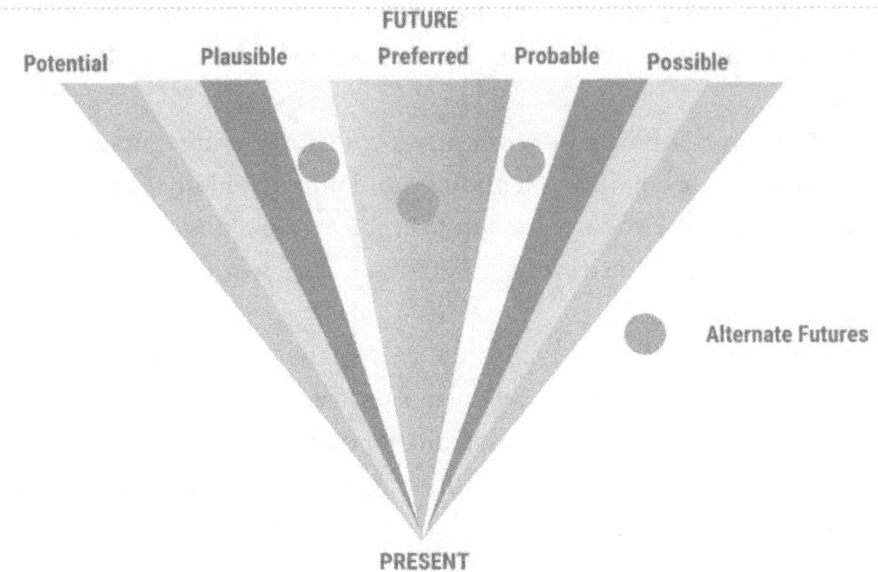

Figure 6-4. The Futures Cone (Adapted). After Hancock & Bezold (1994) and Joseph Voros, "The Futures Cone, Use and History" (Latest Rev. 2017)

The Futures Cone is a heuristic showing sets of futures that widen from the present: Possible (everything consistent with known constraints), Plausible (credible given what we currently understand about how the world works), Probable/Expected (what is most likely if current trajectories continue), and Preferable (value-laden futures a stakeholder group would like to bring about). "Preferable" is not a forecasted set; it cuts across the others and must be argued and negotiated.

Version note: This book uses the Voros (2017) nomenclature (Possible, Plausible, Probable/Expected, Preferable). Some variants add Preposterous (outside Possible) or swap "Probable" for "Projected/Expected." Where the book references the cone later (e.g., Chapter 9), it follows this same version for consistency.

The different segments of the cone represent future states ranging from

- Probable (Expected): The future most likely to occur if current trajectories persist.

- Plausible: Futures that could happen given what we currently judge as causally credible.

- Possible: Futures that might happen because they do not violate fundamental constraints (e.g., physics, hard legal/biological limits).

- Preferable: Futures stakeholders desire; defined by transparent criteria and trade-offs, then assessed for overlap with the plausible/probable sets.

This categorization follows the Futures Cone literature (Hancock & Bezold 1994; Voros 2017), which distinguishes possible, plausible, probable, and preferable futures (often also including projected and wildcard variants).

Developing narratives across this cone of plausibility is a way for futures thinkers to rehearse the wide terrain of contexts an issue could unfold within. For the BCI example, we may envision these:

A Probable future of ubiquitous neural interfaces enhancing productivity and entertainment by 2040.

A Plausible future of a popularist backlash limiting BCI development over privacy concerns.

A Possible future of a BCI-enabled collective consciousness emerging by 2070.

A Wildcard future of a misuse incident spurring an outright BCI ban and neo-Luddite movement.

By sketching out scenarios across this vector of uncertainties, the cone aids in mapping blind spots and considering future branching paths beyond simplistic best/worst case binaries. It fosters more nuanced thinking about the range of possible trajectories that could spin out from emerging issues.

In summary, using frameworks like STEEP+V and the Futures Cone in tandem helps cultivate the broad, systems-level perspective required to navigate the complexities of the future. They equip us with conceptual tools for avoiding dogmatic predictions and accounting for the multitudes of interlinked forces that could reshape the world in transformative ways yet to come.

6.2.2 Identifying and Tracing Signals of the Future

At the center of futures thinking is the practice of detecting and making sense of weak signals—the earliest inceptions of issues, ideas, or dynamics that could grow into future revolutionizing forces. By attuning ourselves to these possibilities, we can get ahead of emerging disruptions and orient ourselves toward preferable trajectories before they fully crystallize.

The signal recognition process begins with immersing ourselves in the traditional array of rich information flows—scanning scientific journals, tracking innovative startups, monitoring policy debates, and keeping our attention focused on nascent cultural shifts and social murmurings. We have to cultivate a futures-conscious perspective, training our minds to pierce through the banalities of day-to-day affairs and detect novel signals wavering between the noise.

For clarity, the items below are grouped by maturity—week signals (early ripples) and maturing signals (watch) to void overclaiming. Some of the early ripples I have traced that could coalesce into future trends include:

- Maturing signals (watch): The expansion of solarpunk from a niche aesthetic into applied initiatives—mutual aid infrastructure, community energy co-ops, repair/permaculture clubs, now informing regenerative design briefs and narratives.

CHAPTER 6   FOUNDATIONAL METHODOLOGIES IN FUTURES THINKING

- Advances in bioengineering like genetic chimeras, xenobiology and biological AI that could obliterate our current ontological boundaries between the natural and artificial systems. How might this upend our philosophies, ethics, and legal systems premised on human/nature divides?

- The convergence of virtual reality, haptic interfaces, and neural mapping that foreshadows a deeper integration of our minds into simulated spatial data environments. What societal transformations could unfold as our phenomenological locus transcends the physical world?

Once identified, the process of tracing signals involves mapping out their lineages—their historical antecedents, contemporary activities accelerating their momentum, and harbingers allowing us to forecast potential future impacts and cascades. We analyze their intersections with other emerging issues, drivers propelling their progression, and prospective cross-pollinations generating new emerging possibilities.

By applying the STEEP+V model here, we can develop multi-dimensional perspectives on the varied facets a signal could trigger across different sectors as it strengthens and evolves. We can also employ tools like Backcasting, Causal Layered Analysis, and the Futures Cone to envision and narrativize different trajectories along which the signal could propagate.

For example, tracking the biodata retrieval and NeuroTech threads, we may map out future scenarios ranging from

A plausible future of a global NeuroTech industry arising by 2040 enabling peak mental performance, hyper-immersive experiences, and even cognitive blockchain economies.

A possible future flowering of "Singularity societies" by 2070 as humans transcend biological limitations and fuse with superintelligent AI systems.

A wildcard future of a cataclysmic "mind hacking" event leading to a popular rejection of neural tech and a move to legally prohibit cognitive enhancement.

The key in all this signal monitoring and futures mapping work is to avoid the simplistic futurism traps of overly linear thinking, dogmatic predictions, or hype-driven technological determinism. We need to hold the tension of seeing emerging possibilities with clarity and empirical rigor, while cultivating a sense of humility about the near-infinite permutations of paths the actual future could take.

6.2.3 The Role of Speculative Design in Shaping Futures Scenarios

While environmental scanning, signals analysis, and frameworks like STEEP+V and the Futures Cone provide crucial context mapping of the future's possibility spaces, speculative design offers a powerful complementary approach for immersing us experientially in envisioned scenarios before they fully materialize.

At its core, speculative design is the intentional crafting of prototypes, products, services, and experiences embodying different future worlds, value systems, and technological capabilities that we wish to critically interrogate. By manifesting speculative artifacts from the future into the present, this practice allows us to confront possible tomorrows in visceral, tangible ways that push beyond mere written scenarios or conceptual models.

In my own futures design work, I have found the act of prototyping future concepts to be an invaluable mode of inquiry for surfacing hidden assumptions, values, tensions, and second-order implications that often remain invisible when speculating on an abstracted plane. The very process of bodging together rudimentary simulations demands that we wrestle with the granular details and systemic interactions required for a particular envisioned possibility to cohere into a plausible future reality.

For example, a few years ago I created an independent project, prototyping a future ecosystem of a smart home product, with a cooking device, envisioned around the principle of "congruous design"—an approach prioritizing technologies that seamlessly integrate into our daily life rituals and practices rather than imposing themselves as disruptive external efficiencies.

While the premise sounded elegant in theory, crafting low-fi manifestos of this domestic ecology revealed deeper tensions we had to wrestle with. How might interactive surfaces become dynamically expressed into architectural environments without compromising functionality? What if individual needs shifted fluidly between public and private contexts—how could interfaces adapt? If domestic labor still existed in this future, how might the technologies preserve human qualities of care?

Navigating these experiential questions pushed us to confront complex social, ethical, and even spiritual friction points that would need to be resolved for this imagined future to harmonize with our core human values and ways of being.

This type of immersive, artifact-driven process not only surfaces unseen tensions to wrestle with, but also sparks profound insights about how the future could be alternatively designed for different value prioritizations and systemic outcomes. In another futures project exploring sustainable food ecologies for the year 2040, we created a simulation prototype of three wildly divergent future scenarios:

Urbanika: A closed-loop urban vertical farming ecosystem powered by AI-optimized microbe cultures that dynamically grow complete nutritional solutions based on individual metabolic profiles uploaded to the system.

Terran Revival: A return to bioregional, hunter-gatherer-cultivator lifestyles enabled by engineered bio-wildernesses abundant with hybrid flora/fauna and DIY genetic mashup kits for reseeding ecosystems.

Solarpsilon: A post-scarcity future of in-home nano-manufacturing pods that digitally compile any food desire from radical energy-to-matter conversion technology powered by the sun.

Manifesting these varied visions into sketches, models, and experiential sketches illuminated the radically different value propositions, narratives of progress, and socio-political dynamics undergirding each scenario's logic. The experiential artifacts almost became Rorschach prompts for us and subject-area experts to grapple with how our fundamental worldviews could take shape in material culture.

Beyond enriching our scenario sensemaking, speculative design also enables us to socialize futures in ways that spur broader engagement and imaginative participation. When communities, companies, or policymakers are presented with the tangibility of a future made imminent, it catalyzes richer dialogue, debate, and collaborative evolution of preferred trajectories.

By merging artifacts into the lived present, speculative design collapses temporal distances and makes potential futures viscerally appreciable. This unlocks futures thinking from the confines of the imagination alone and propels it into an embodied, participatory phenomenon for collectively prototyping the worlds we wish to inhabit. When united with futures scanning, signals research and scenario planning, speculative design equips us with a potent integrated toolkit for manifesting the futures we deem most vital into reality.

6.3 Summary

Futures design and speculative design are creative approaches that focus on making potential future scenarios tangible through narratives, prototypes, and artifacts. Unlike traditional data-driven forecasting, these methods emphasize imagination and storytelling to immerse people in possible futures, sparking inquiry and inspiring change. By materializing these futures in the present, futures design helps us explore the implications of emerging technologies and societal shifts in a visceral, experiential way. For example, projects like the Biolitt habitat and speculative artifacts like those depicted in the film *Her* illustrate how design can transcend current paradigms and engage people in meaningful discourse about the future.

These techniques are particularly powerful in strategic contexts, where they enable envisioning exercises and the development of experience prototypes. Through speculative design, participants can inhabit and explore potential futures, leading to deeper collective understanding and accelerated innovation. Whether used to catalyze discussion or to refine strategic visions, futures design and speculative design empower us to actively shape the trajectory of our collective future by aligning it with our values and intentions.

CHAPTER 7

Tools and Techniques for Futures Thinking Exploration

7.1 Scenario Development and Analysis

Scenarios are powerful tools for exploring alternative future possibilities and their implications. By constructing and analyzing different scenarios, we can better anticipate potential changes, identify emerging issues, and prepare more robust strategies. As illustrated in Figure 7-1, scenario development is a core methodology in futures thinking that helps organizations and decision-makers avoid cognitive biases like straight-line extrapolation of current trends.

CHAPTER 7   TOOLS AND TECHNIQUES FOR FUTURES THINKING EXPLORATION

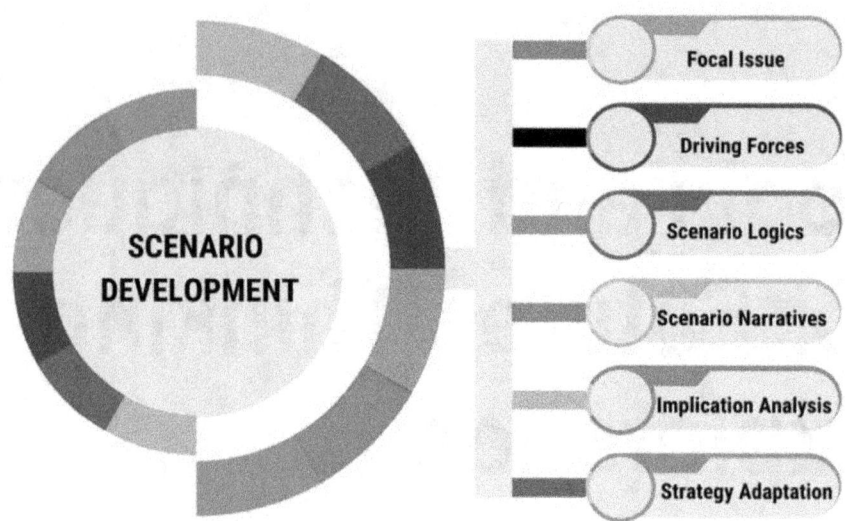

Figure 7-1. *Scenario Development and Analysis*

Scenarios are powerful tools for exploring alternative future possibilities and their implications. By constructing and analyzing different scenarios, we can better anticipate potential changes, identify emerging issues, and prepare more robust strategies. Scenario development is a core methodology in futures thinking that helps organizations and decision-makers avoid cognitive biases like straight-line extrapolation of current trends.

At its core, a scenario is a coherent narrative description of a possible future reality that could emerge from the interplay of various driving forces, events, and dynamics. Scenarios are not meant to be predictions, but rather plausible, pertinent, and challenging accounts of how the future may unfold differently than expected. The value lies in expanding understanding of key uncertainties and discontinuities that could reshape the big picture.

The scenario development process typically involves several key phases (adapted from classic scenario practice: Schwartz 1991; van der Heijden 1996; Schoemaker 1995; Chermack 2011):

CHAPTER 7   TOOLS AND TECHNIQUES FOR FUTURES THINKING EXPLORATION

1. Defining the focal issue or decision to be explored: This scopes the terrain and establishes boundaries for the scenario work.

2. Identifying driving forces, predetermined elements, and critical uncertainties: This phase surfaces major trends, volatile factors, emerging issues, and potential ruptures in the broader landscape relevant to the focal issue.

3. Selecting scenario logics or "axes of uncertainty:" Based on the driving forces identified, a few overarching dimensions of uncertainty are chosen to structure divergent scenarios around (e.g., a scenario with high economic growth vs. stagnation).

4. Fleshing out scenario narratives and paths: For each scenario framework, rich descriptive narratives are built out describing the events, dynamics, impacts, and outcomes that could occur as that scenario trajectory takes shape over time.

5. Implications analysis: Each scenario narrative is analyzed to explore its consequences, challenges, and strategic implications for the organization or issue at hand. This allows rehearsing potential futures in advance.

6. Decision support and strategy adaptation: Based on the scenarios' implications, strategies and plans can be stress-tested, updated, or reformulated to be more robust across a range of scenarios.

Effective scenario development requires involving diverse perspectives and combining both analytical rigor and creative narration. Methods like environmental scanning, cross-impact analysis, and interviewing

CHAPTER 7   TOOLS AND TECHNIQUES FOR FUTURES THINKING EXPLORATION

subject matter experts all play a role in grounding scenarios in substantive research. At the same time, scenarios must inspire new thinking by challenging conventional wisdom and mental models.

When done well, scenario work bypasses human tendencies toward overconfidence and tunnel vision about the future. It equips organizations and leaders to perceive emerging risks and opportunities earlier, while fostering more agile, resilient strategies attuned to fundamental uncertainties and discontinuities. In an era of accelerating change, scenario analysis is an essential capacity for futures literacy.

Here is an example to illustrate the scenario development process outlined above with our simulated XRAI company that builds electric vehicles:

Let's assume the focal issue is exploring the future of electric vehicle (EV) adoption and its impacts on a major automaker over the next 15 years.

1. Defining the focal question: How might dynamics shaping EV adoption unfold, and what are the strategic implications for XRAI's product roadmap, manufacturing, and supplier networks by 2035?

2. Identifying driving forces:

    - Predetermined elements like climate change, urbanization trends, traditional automaker transitions

    - Major uncertainties like speed of battery technology improvements, shifts in consumer preferences, government policies incentivizing EVs, electricity grid upgrades

CHAPTER 7   TOOLS AND TECHNIQUES FOR FUTURES THINKING EXPLORATION

3. Selecting scenario axes: After analysis, two overarching uncertainties are chosen as scenario axes: A) Degree of policy/regulatory support for EVs (high vs. low) B) Pace of EV battery cost declines (fast vs. slow)

4. Fleshing out scenarios: Four scenarios are constructed at the intersection of these axes: A) EV Transformation—High policy push and fast battery cost declines lead to widespread EV adoption B) Niche Disruption—Low policy support but fast battery breakthroughs enable EVs to disrupt certain urban/premium segments C) Gradual Evolution—Moderate, market-driven EV growth due to slow battery improvements despite policy support D) ICE Age Continued—Internal combustion engines remain dominant with slow battery progress and low EV incentives

   Rich narratives describe how production scales, supply chains, consumer demand, energy systems, and competitive landscapes could co-evolve differently in each scenario from 2020 to 2035.

5. Implications analysis: Each scenario's impacts on Company X are analyzed—e.g., In the EV Transformation scenario, what capabilities and supplier relationships would need to be reconfigured? How could the Gradual Evolution scenario strand manufacturing assets?

6. Strategy adaptation: Based on the analysis, Company X revised its product strategy with a dual-track approach—rapidly scaling EV lines to

be positioned for an EV Transformation, while hedging by continuing to invest in optimizing internal combustion engines for a slower transition. Supply chain resilience and partnerships were also prioritized.

By working through this scenario process, the company avoided simply extrapolating current EV demand. Instead, it perceived the major uncertainties and developed a more adaptive, resilient strategy across multiple scenarios.

What Changed Because of Scenarios (Decision Impacts)

The exercise led to several decisions that would likely not have surfaced from a straight line EV demand forecast:

Dual options with triggers: Approved two strategy options—EV acceleration and legacy bridge, each with quantitative trigger points (e.g., > 42% regional EV adoption or <$80/kWh battery pack price for three consecutive quarters).

Portfolio hedge: Rebalanced the product roadmap to 60% EV/40% ICE over 24 months, with quarterly review against the triggers above.

Supply chain resilience: Added a secondary cathode supplier and regionalized final assembly in two markets to reduce geopolitical and logistics risk identified in the "Stagnation/Fragmentation" scenario.

Learning bets: Funded three small experiments (V2G pilots with a utility, LFP chemistry trial, and a swap station partnership) explicitly tied to the critical uncertainties surfaced in the scenarios.

KPIs and Monitoring: Established a signal dashboard (policy incentives, charging density, chip lead times, consumer segment shifts) and linked it to monthly product council meetings.

Governance: Created a Scenario Review cadence (semiannual) to re-test assumptions, retire scenarios, and add new ones as signals shift.

Rationale: These actions translate scenario insights into options, triggers, and staged investments, making the strategy adaptive rather than a single bet.

7.1.1 Creating and Analyzing Scenarios Using the Futures Wheel and Futures Cone

Two powerful techniques for developing and exploring scenarios are the Futures Wheel and the Futures Cone. When combined, they provide a structured yet flexible approach to mapping out alternative future trajectories and their branching implications.

The Futures Wheel is particularly valuable for fleshing out the inner logic and ripple effects within a given scenario narrative. To construct a Futures Wheel for scenario development (Glenn's Futures Wheel method; see Glenn 1972; Millennium Project FRM 3.0):

1. Define the central scenario driving force, event, or focal issue to be explored (e.g., widespread adoption of autonomous vehicles).

2. Write this clearly in the center of the wheel diagram.

3. Identify and map the primary/direct impacts or consequences radiating outward from this central premise using short spokes.

4. For each primary impact, identify the secondary impacts it could trigger, branching outward with new spokes.

5. Continue expanding the wheel with higher-order impacts and consequences fanning outwards in rings.

CHAPTER 7   TOOLS AND TECHNIQUES FOR FUTURES THINKING EXPLORATION

This systematic mapping helps expose the cascading implications of the central scenario driver across different sectors and dimensions—social, technological, economic, political, environmental, etc. It makes the scenario richer by tracing out its interdependencies.

For example, a Futures Wheel exploring an "Autonomous Vehicles Disruption" scenario could map impacts like

- Primary: Reduced automobile ownership, transportation costs plummet

- Secondary: Vehicle insurance industry disrupted, urban land use shifts

- Tertiary: New mobility service business models, vehicle communications infrastructure

Once constructed, the dense visualized web of impacts sparks new insights about critical issues, tensions, and intervention points within that scenario. Contradictions and feedback loops may also surface.

The Futures Cone complements this by providing a high-level framework for developing a set of contrasting, cogent scenarios around a focal issue or axis of uncertainty. The cone metaphor depicts

1. The current reality as a semi-solid present

2. An expanding aperture of alternative plausible futures fanning outward over time, ranging from

    - The Expected/Probable future based on persisting trends (the Cone's central axis)

    - Outlier Possible and Wildcard scenarios (the outer edges of the Cone)

By selecting 3–4 scenarios' positions mapped along different trajectories within this Cone of Plausibility, you can develop narratives exploring fundamentally different future states and dynamics, such as

CHAPTER 7 TOOLS AND TECHNIQUES FOR FUTURES THINKING EXPLORATION

- A Probable "Green Autonomous Mobility" scenario where sustainability drives AV adoption
- A Wildcard "TransportChannel Sim-Life" of virtual reality replacing physical transit
- A Possible "MobilityApart" where inequality divides access to automated transportation

Once constructed using techniques like the Futures Wheel, each scenario narrative can be comparatively analyzed for impacts, strategic openings, and early indicators signaling its emergence.

In tandem, the Futures Wheel surfaces nuances within individual scenarios, while the Futures Cone provides an expansive framing of the overall possibility space. Their contrasting yet complementary perspectives enhance scenario development and implication analysis. Used iteratively, they build richer narratives accounting for both granular and structural uncertainties—catalyzing more robust strategic preparedness.

Let's bring this context in perspective. Here is a more detailed example of using the Futures Wheel and Futures Cone methods together for scenario development and analysis as illustrated in Figure 7-2:

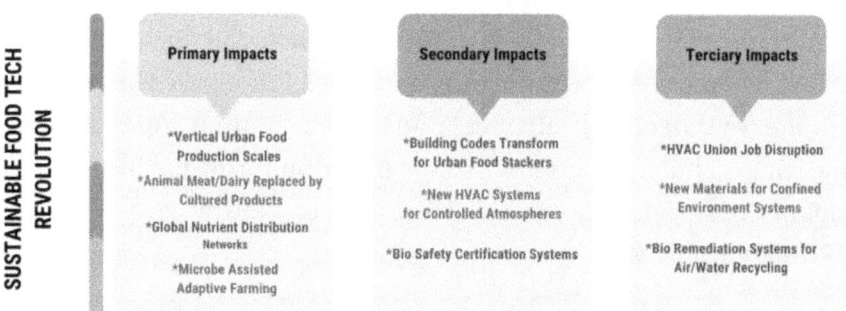

Figure 7-2. Futures Wheel/Futures Cone Example

CHAPTER 7   TOOLS AND TECHNIQUES FOR FUTURES THINKING EXPLORATION

Let's take the focal issue of exploring the future of food systems and food security in 2040.

First, we could use the Futures Cone to map out a set of contrasting scenarios along two major axes of uncertainty:

Axis 1: Degree of ecological disruption from climate change (ranging from Catastrophic to Manageable)

Axis 2: Technological transformations in food production systems (ranging from Radical/Disruptive to Incremental)

Plotting these axes creates four quadrant scenarios at the cone's outer edges:

1. Sustainable FoodTech Revolution (Manageable Climate/Radical Tech)

2. Climate Food Crisis (Catastrophic Climate/Incremental Tech)

3. Bio-Integrated NutriCultures (Manageable Climate/Radical Tech)

4. Agro-Ecological Descent (Catastrophic Climate/Incremental Tech)

We could then use the Futures Wheel to map out the impacts and dynamics within each of these narratives. Let's take the "Sustainable FoodTech Revolution" scenario and construct a Futures Wheel for it:

[Center of Wheel]: Sustainable FoodTech Revolution—Precision bio-engineering, cellular agriculture, and digital monitoring enable a radical abundance of sustainably produced nutrition globally by 2040.

[Primary Impacts]:

- Vertical urban food production scales
- Animal meat/dairy replaced by cultured products
- Global nutrient distribution networks
- Microbe-assisted adaptive farming

CHAPTER 7   TOOLS AND TECHNIQUES FOR FUTURES THINKING EXPLORATION

[Secondary Impacts for Urban Food Production]:

- Building codes transform for urban food stackers
- New HVAC systems for controlled atmospheres
- Bio-safety certification systems

[Tertiary Impacts for New HVAC Systems]:

- HVAC union job disruption
- Novel materials science for confined environment systems
- Bio-remediation systems for air/water recycling

And so on, with each primary impact spawning its own branches of secondary and tertiary resonances mapped outward in the wheel's concentric rings.

We could construct similar wheels for each of the other cone scenarios, surfacing unique impact clusters like the following:

Climate Food Crisis:

- Famine belt expands globally
- Massive human migration patterns
- Food noir market violence
- National food hoarding and protectionism

Bio-Integrated Nutrition Cultures:

- Cultural backlash to radical nutrient forms
- Global calorie corporations
- Open source nutrient editing communities
- Bio-Hactivist movements

CHAPTER 7   TOOLS AND TECHNIQUES FOR FUTURES THINKING EXPLORATION

This makes the scenarios richer and more multidimensional. We may realize some scenarios have impact clusters making them more unstable, less resilient, or problematic—like the Climate Food Crisis wheel exposing major national/global security issues.

We could then analyze the scenarios comparatively based on

- How expansive and dense their Futures Wheels are (scope of impacts)

- What critical issues, tensions, or leverage points surface when mapping their impacts

- Whether the scenario impacts validate or undermine the scenario's core premise

- What potential wildcards or cross-scenario impacts exist between wheels

This combined Futures Cone/Wheel analysis makes the scenarios more tangible and holistic. It expands our perception of each scenario's dynamics while also stress-testing their viability against potential impact pathways. We build preparedness for navigating the scenarios through forethought.

In the analysis, we may realize one of the cone scenarios is actually unlikely to persist based on predictable impact dominoes revealed by its Futures Wheel. Or we may decide to hybridize elements from different scenario wheels into a new, more resilient vision integrating the most desirable and coherent impacts. The point is that iterating between the future scenario logics (the Cone) and the granular impact mapping (the Wheel) allows us to create richer, more challenging and adaptive scenario narratives and strategy stress-testing. It's a powerful tandem approach for evolving our futures, preparedness, and wisdom.

CHAPTER 7   TOOLS AND TECHNIQUES FOR FUTURES THINKING EXPLORATION

7.1.2 Applying Jim Dator's Four Futures Framework

When developing scenarios and exploring potential future trajectories, it's critical that we avoid the classic pitfall of getting lulled into a singular vision—whether utopian or apocalyptic. The future is radically plural, with divergent possibilities pulling in multiple directions based on the complex interplay of forces and human choices. That is why I always push back against any futures work pedaling just one predetermined future, no matter how vivid or compelling the storytelling. We have to stretch our perception to encompass a wider terrain of alternatives.

This pluralistic perspective is where the late Jim Dator's concept of generic alternative futures comes in. Rather than fixating on one potential scenario, Dator outlined four foundational narrative categories that any serious foresight work should consider—a framework for challenging our assumptions and expanding our futures perception.

At its core, Dator's four futures provide archetypal storylines representing fundamentally different societal premises and trajectories. They are as follows:

1. The Continuation future: This is essentially the path of least resistance, a straight-line extrapolation of present trends, incremental change, and the perpetuation of our current global operating system and worldview. It's the "business-as-usual" scenario that most strategic plans implicitly bet on.

2. The Disciplined future: This archetype envisions a tightening of institutional order, control, and top-down management of societal affairs, whether through centralized governance structures, security/militarization, fundamentalist ideologies, or totalizing technologies. It's the "Big Brother" future we've been warned about.

3. The Transformational future: On the opposite end is this archetype of radical, disruptive, grassroots-driven change that transcends and supplants the status quo entirely. This could include post-scarcity socioeconomic new orders, transhumanist evolution, indigenous revivalisms, or histrionic shifts in human consciousness and being.

4. The Collapse future: As the name implies, this is the nightmare archetype of civilizational decline, systems failure, resource depletion, conflict, and potential existential catastrophe—the apocalyptic scenario that prepper movements insure against.

Now, to be clear, these are broad narrative buckets representing vastly different societal trajectories and premises, not hardline predictions. The power of Dator's framework is in using these distinct archetypes as scenario lenses for perceiving how different choices, driver forces, and failures could reshape the future in radically different ways.

For example, when exploring the future of governance models, we could construct distinct scenarios representing each of these narrative premises:

- A Continuation scenario of nation-state incrementalism and muddling through

- A Disciplined scenario of rising securitization and authoritarian structures

- A Transformational scenario of emergent self-organizing systems and biomimicry governance

- A Collapse scenario where climate/resource pressures fracture governance capacity

Deliberately constructing scenarios through each of these lenses forces us to map out the compounding impacts and contingencies under those divergent trajectories. It makes us confront how different the future could be based on the roads we choose (or don't). Applying this archetypal framing reveals blind spots in our expectations and strategic vulnerabilities across multiple futures.

In my own foresight work, I often deploy Dator's framing alongside other scenario mapping methods like the Futures Cone and Futures Wheel. I'll use the four archetypes to define distinct scenario trajectories within the possibility cone, then construct rich wheels diagramming the nuanced impacts and dynamics within each archetypal path. This integration of frameworks allows me to create both structurally divergent yet granular and cogent scenarios for almost any focal issue I'm exploring—whether the future of food systems, human machine symbiosis, or global consciousness.

7.1.3 Building Detailed Future Scenarios and World Building Techniques

Once we have constructed an initial framework for a set of future scenarios using methods like the Futures Cone and Wheel, the next phase is fleshing out those scenarios into rich, immersive narrative worlds. This world-building process helps make the scenarios tangible and visceral rather than abstract thought experiments.

At this stage, we move beyond just mapping out broad trajectory pathways and impact dynamics. We now focus on bringing the subtle textures and experiential qualities of each potential future to life through vivid descriptive details. The aim is to make each scenario palpable—evoking the look, feel, social realities, and sociocultural ambiances that could emerge.

CHAPTER 7   TOOLS AND TECHNIQUES FOR FUTURES THINKING EXPLORATION

One powerful technique is developing multimedia artifacts that manifest these future worlds. We construct speculative artifacts like design prototypes, science-fiction storytelling, virtual reality experiences, and multimedia simulations that give participants an immersive, interactive glimpse into how the different scenarios could manifest.

For example, if exploring scenarios around the future of human–machine symbiosis, we could develop

- A narrated clip featuring a family using seamless brain–computer interfaces in their smart home
- Rendered models of neural implant devices with didactic schematics explaining the technology
- An AI virtual assistant persona that showcases the extended cognition capacities
- A multisensory VR experience simulating an enhanced state of mind uploading skills on demand

Such artifacts make the often-ambiguous contours of scenarios visceral and tangible. They allow people to inhabit potential realities rather than just intellectualizing them. This immersive quality sparks richer discussions and enhances prepared-mind pattern recognition for discerning between emerging realities.

Another powerful world-building approach is developing a comprehensive description of the scenario's ambiance and experiential qualities. This involves comprehensively articulating details like the following:

Sociocultural Fabric:

- Prevalent values, beliefs, and narratives circulating
- Shifts in norms, ethics, and social dynamics
- How identity, community, and relationships are structured

CHAPTER 7 TOOLS AND TECHNIQUES FOR FUTURES THINKING EXPLORATION

Built Environments:

- Architecture and engineered spaces
- Ambient media/computing landscapes
- Design languages and aesthetic sensibilities

Organizational Ecosystems:

- Dominant power structures and governing logics
- Prevalent business/economic models
- New industries, institutions, and human geographies

Environmental Dimensions:

- Natural ecosystem health and human–nature interfaces
- Predominant forms of energy, food, and production systems
- Climate conditions and resource flows

By saturating scenarios with this level of immersive, multidimensional detail, we make them as real and visceral to participants as the world they experience today. We bypass typical failures of imagination and ignite the experiential foresight required to comprehend—and effectively navigate—radically divergent planetary trajectories.

These world-building techniques can be combined with other methods from this toolkit as well. For example, exploring wildcard scenarios using something like Causal Layered Analysis to unpack prominent metaphors and worldviews. Or applying systems-modeling to simulate complex dynamical pathways within a particular scenario reality.

The key is taking the time to construct scenarios that are not just conceptual narratives, but fully experienceable potential worlds that awaken us to longer-term contingencies. By enveloping people in immersive futures before they unfold, we enhance societal futures literacy and capacity for creating regenerative paths forward.

CHAPTER 7   TOOLS AND TECHNIQUES FOR FUTURES THINKING EXPLORATION

Fortunately, emerging AI capabilities are opening up powerful new avenues for constructing vivid, multisensory scenario worlds. Generative AI systems can be applied throughout the world-building process to accelerate and enhance our ability to manifest future realities before they manifest.

For example, we could use language models to rapidly generate first-draft descriptions, narratives, and artifact design specifications across all the dimensions listed above—sociocultural, built environments, etc. Rather than scenario teams laboriously drafting every detail, we could provide high-level outlines and premises, then instruct AI to "hallucinate" initial multisensory rendered worlds based on those outlines using its multimodal knowledge bases.

From there, human participants could intuitively edit, iterate, and flesh out the autogenerated scenario worlds by simply instructing the AI system until a desired level of descriptive fidelity is reached. This AI-assisted co-creation workflow could dramatically accelerate our ability to construct experiential scenario textures, especially for more speculative contexts pushing the limits of human imagination.

We could even use advanced text-to-image generative AI to rapidly visualize artifacts from the scenario worlds—including artifacts that don't yet exist like future technologies, living environments, cultural objects, and more. This enhances our ability to spark immersive experiences and make the scenarios viscerally resonant. Imagine being able to iteratively generate photorealistic visualizations of different scenario world dynamics until they really "feel" coherent and plausible.

For full embodied immersion, generative AI could also be combined with virtual reality, augmented reality, haptics, and other extended reality technologies. Rather than just static descriptions, we could develop AI-powered simulations allowing people to experientially inhabit and navigate the different scenario worlds in rich XR environments. This immersion cultivates deeper felt-sense resonance with how vastly different planetary civilizational trajectories could subjectively manifest.

CHAPTER 7   TOOLS AND TECHNIQUES FOR FUTURES THINKING EXPLORATION

For instance, in our food systems scenario example, we could develop an AI-generated open-world simulation allowing participants to experientially explore and inhabit the different urban food landscapes, experience future food rituals and social dynamics, or walk through speculative food production facilities. Such embodied journeys allow people to get a visceral felt-sense for the divergent human/environmental realities inherent to each scenario.

However, we must apply critical wisdom in leveraging AI and XR for scenario world-building. There are risks like AI hallucination leading to incoherent or nonsensical futures if not carefully filtered. Or AI systems amplifying present-day cultural biases and blind spots into visions of the future. With XR immersion, we must navigate issues like virtual realities inducing experiential attachment or distress depending on the qualities depicted.

As such, any use of AI and XR should be a supplement to rigorous human insight, contextual research, and intentional curation by foresight practitioners attuned to deeper metaphors and civilizational currents. We must see these as co-creative tools to spark intuition and depict grounded pluralistic possibilities, not as autonomous oracles generating singular visions of the future.

7.2 Imagining and Prototyping Futures

While techniques like scenario planning allow us to explore alternative future trajectories, there is also immense power in envisioning and giving form to the futures we actually want to create—the worlds we would deliberately aim to transition toward if we could cocreate them by design.

This is the domain of imagining and prototyping desirable futures—a critical practice for transcending path dependencies and actualizing radically different planetary cultures and civilizational operating systems aligned with our highest aspirations and values.

CHAPTER 7   TOOLS AND TECHNIQUES FOR FUTURES THINKING EXPLORATION

At its core, this work invites us to engage in extended acts of imagination and possibility modeling. We temporarily suspend assumptions about what's probable or seemingly inevitable. Instead, we allow ourselves to radically reimagine human societies, technologies, environments, and ways of being from a fundamentally new grounding in what's optimally desirable and regenerative.

This is not an exercise in static utopian fantasies, but in developing pragmatic, livable visions grounded in ethics, science, and human wisdom. The aim is to manifest actionable, participatory simulations of preferred futures that can inspire and provide a beacon for navigating our way toward their realization through infinite incremental steps.

There are many approaches and techniques for doing this visionary work, but they generally involve the following:

Reframing Fundamental Premises:

We step back and reexamine the core axioms, values, and framing metaphors underlying modern human civilization—our relationships to self, society, technology, Earth. We shed inherited paradigms and redefine first principles from which to reimagine human systems and trajectories.

Mapping Possibility Spaces:

With new grounding premises, we explore the possibilities of what could be, modeling vastly alternative sociotechnical systems, cosmologies, lifeways, and modes of being that actualize our highest aspirations and deepest wisdom as a species.

Prototyping Future Artifacts:

To make these reimagined futures tangible and participatory, we construct experiential prototypes—speculative artifacts, interactive models, and simulations that allow people to inhabit and experience the visceral textures of thriving in these alternative futures.

Community Visioning:

This work happens through transdisciplinary collaboration and diverse community dialogue. We engage in multimedia storytelling, worldbuilding, and participatory prototyping to collectively envision and uplift the pluralistic yet universal human futures we wish to create.

Over the coming sections, we will unpack specific techniques for reframing premises, mapping possibilities, prototyping artifacts, and uplifting community visions of thriving planetary futures. But first, it's important to center ourselves on why this intentional work matters.

The primary motivation is cultivating human agency over our long-term trajectory as a species. For too long, we have been trapped in cycles of path dependency and reactivity—future scenarios being shaped more by inertia, power dynamics, and perceived inevitabilities than conscious, ethical choice. This has led to civilizational systems and trajectories wildly out of alignment with our fundamental human potentials.

By reclaiming our ability to envision and prototype futures by design, we disrupt these cycles and embody our power as cocreators in shaping the realities we will transition toward. We awaken to a deeper, participatory mode of evolution—one where our grandest imaginations and values aren't just abstract ideals, but tangible simulations we can actively inhabit and cocreate into being. We will explore specific tools and practices for doing this powerful work next.

7.2.1 Radical Reimagining and Innovation for Desirable Futures

Context: This section provides the mindset that underpins the tools in Chapters 6–7; it's the bridge from imagination to decision.

The first step in envisioning and prototyping the futures we actually want to create is reframing the fundamental premises and grounding metaphors from which we construct human realities. Too often our visions of the future remain confined by implicit axioms and paradigms inherited from the past.

CHAPTER 7   TOOLS AND TECHNIQUES FOR FUTURES THINKING EXPLORATION

To transcend these blinders, we can engage in radical acts of reimagination—suspending ingrained assumptions and redefining first principles from which to innovate fundamentally new human narratives, systems, and trajectories aligned with our highest potentials.

This work starts with reexamining the core metaphysical, epistemological, and ethical premises underlying modern civilization. We must interrogate the root axioms and framing stories shaping our dominant human lifeways, economies, technologies, and relationships to self and world.

For example, we may need to redefine premises like

- The nature of human identity, consciousness, and ways of being
- Definitions and axioms shaping "progress" and "development"
- Premises about the human/nature relationship and ecological order
- Axioms around scarcity, resources, and economic valuation
- Narratives about time, change, the nature of the future itself

By surfacing and reexamining these root metaphysical narratives, we open up possibility spaces for radical reimagination. We can shed paradigms confining us to regressive and anthropocentric premises, embracing grounding stories that harmonize human self-actualization with ecological flourishing.

Emerging wisdoms from fields like consciousness studies, indigenous cosmologies, future sciences, and evolutionary anthropology can provide fertile ground for redefining these first principles. We draw from expansive

CHAPTER 7   TOOLS AND TECHNIQUES FOR FUTURES THINKING EXPLORATION

truths about interconnectedness, temporality, the nature of being, and mind to construct new narratives about what it means to be human and to become a thriving planetary civilization.

From this space of reframed metaphysics, we can then engage in radical innovation—prototyping alternative models and systems that actualize these new premises into livable futures. This may involve reimagining everything from

Human Lifeway Systems: New models for housing, community, sustenance, healing, and lifecycle structures that transcend industrial norms. Evolution from consumer cultures to regenerative planetary cultures.

Socioeconomic Frameworks: Noncapitalist models encoding regenerative circuit economies, actualized universal abundance, and new relationships between human/technology/nature realms.

Governance and Organizing Logics: Decentralized, self-organizing systems that harmonize agency/accountability across scales. Biocompatible decision-making and scaling processes.

Science and Knowledge Paradigms: Integrating first-person ontologies into scientific methods. Hybridizing analytic and generative knowledge models across domains.

Technology Cosmologies: Re-envisioning technological progress through biomimetic ethics and designing for cosmo-compatibility across physical/digital/metaphysical realms.

At their core, these radical reimaginings are about evolving beyond industrial premises of separation—transcending paradigms that cast humans as discrete entities extracting from external natures. Instead, we develop integral narratives harmonizing human self-actualization with the flourishing of Earth's living systems and cosmic creative potentials.

This expands our capacities to prototype starkly different civilizational trajectories—modes of being, relating, innovating, and creating grounded in premises of interconnected wholeness and regenerative becoming.

Rather than mere techno-utopian visions, we prototype biocultural lifeways that could viably operationalize universal human thriving within the cosmic ecology.

Of course, this level of reimagination and integrated prototype development demands profound transdisciplinary integration and creative synthesis. It's a radical cocreative practice at the fertile intersections of diverse first-person wisdom traditions, material and life sciences, technical disciplines, social change, and future studies.

We work under the premise that when done with intellectual integrity and ethical care, this practice births visions and artifacts that feel vibrantly possible yet expansively liberating—prototypes of desirable futures that attune us to bigger waves of complex, cooperative planetary and cosmic becoming we could actively participate in cocreating.

7.2.2 Developing Future Personas, Locations, and Artifacts

Once we have reframed the metaphysical and ethical premises for reimagining desirable futures, we can engage the work of modeling out the experiential textures of what those futures could actually feel like to inhabit.

A powerful way to make these reimaginings tangible and immersive is by developing multimedia portraits of future personas, locations, and artifact experiences. We construct multimedia prototypes that allow people to viscerally envision the qualitative life worlds and vernacular realities emergent from the new paradigms.

Example. 2032: Heat-Resilient Neighborhood.

Scenario Context: Phoenix metro in a "Hotter, Fragmented" trajectory (frequent 45–48 °C heatwaves, stressed grid, neighborhood level organizing).

CHAPTER 7   TOOLS AND TECHNIQUES FOR FUTURES THINKING EXPLORATION

Signals: Drivers (traceability); Rising wet-bulb days; IRA/local rebates for community solar+batteries; cooling center strain; heat mortality disparities.

Future Persona (Aisha, 34): Community organizer, caregiver; renter; leads a tenant's WhatsApp mutual aid group. Needs: reliable cooling during outages; medication refrigeration; safe night air.

Constraints: Low income, landlord approvals.

Future Location: 1970s apartment block, west-facing units; retrofitted shade sails; stairwell becomes "cool corridor"; roof hosts shared PV+battery.

Speculative Artifact: CoolBlock Kit. Low cost, modular package (phase change cooling panels for a bedroom; battery ready DC fans; door/window gaskets; shared freezer pods for meds).

Use and Learning Goals: Run a 72-hour simulated outage; measure sleep quality, indoor temps, battery draw; gather neighbor feedback on cost/maintainability.

Decision Hook: If kit keeps bedrooms ≤30 °C at night and payback < 4 years (with rebates), the housing association pursues a 50-unit pilot.

Worksheet: Future Persona—Location—Artifact

Scenario name and year: _____

Signals → drivers (3–5): _____

Persona (name/role): goals _____ / constraints _____ / equity notes _____

Location (sketch facts): climate/infra _____ / social context _____

Speculative artifact/service: purpose: _____ / how it works _____ / cost band _____

Experiment: what to test _____ / success criteria (metrics and thresholds) _____

Implications: who benefits/loses _____ / next decision if criteria met

7.2.2.1 Future Personas

One common approach is developing biographical profiles or experiential vignettes offering glimpses into the lived experiences and ways of being for people in the envisioned futures.

For example, if prototyping a future grounded in prosperity circulating economies and skill-sharing guilds, we could develop a future persona like this:

Mika is a 35-year-old experience alchemist and community catalyst living in the Byxbee bioregion. A member of the dream builders guild, Mika helps design immersive play shops blending contemplation, skill-sharing, and metaphysical reflection...

These multimedia persona vignettes, perhaps blending narrative text, ambient audio/visuals, and even AI-rendered digital artifacts, allow people to get a visceral felt-sense for the vernacular lifeways and mindsets that could become normalized in such a future. They make the future tangible and resonant.

7.2.2.2 Future Locations

Similarly, we can develop vivid prototypes and simulations depicting the design, atmosphere, and dynamics of future locations or built environments aligned with the reimagined premises.

Perhaps we are modeling futures where nature, technology, and human habitats are interwoven through biomimetic design principles. We could develop renderings and explorable VR/AR builds depicting

The Sao Maura Bioscape—A floating arcology anchored in the Gulf of Mexico, blending marine permaculture with advanced material science and ambient AI systems. Made from catalytic living architecture, the bioscapes' hyperbolic hydroponic lattices provide housing, sustenance, and energy circulation for its 20,000 inhabitants...

CHAPTER 7 TOOLS AND TECHNIQUES FOR FUTURES THINKING EXPLORATION

These prototypes of future cityscapes and communities transport people into immersive environments texturing how new built environments, technologies, and social/cultural practices could manifest.

7.2.2.3 Future Artifacts

On a more granular level, we can model tactile objects and artifacts that could populate everyday life in the desired futures. This might include prototypes of future technologies, cultural artifacts, art pieces, and more meant to simulate the experience of interfacing with this reimagined world.

If envisioning a future where regenerative nanocomputing transcends today's screen interfaces, we could develop example artifacts like

The Mycolio—A handheld mycelium node interface; the Mycolio acts as a lens into the planet's neural web. By intuitively attuning embodied awareness, it provides ambient information streaming tailored to each person's interests or quests...

We could 3D print models or prototype immersive AR/VR/MR (Augmented Reality/Virtual Reality/Mixed Reality as illustrated in Figure 7-3) interactions using tools like Shapes XR, around such speculative artifacts as a way to simulate the material and phenomenological experience of being in that future reality.

These multimedia portraits, whether textual vignettes, rendered environments, objects, or immersive simulations, are about making the reimagined future resonant through felt-sense experience versus abstract speculation. They spark intuitive insights into the ontological and phenomenological qualities of vastly alternative future realities.

Figure 7-3. Shapes XR Prototyping Tool for VR, MR, and AR

7.2.3 Prototyping Future Concepts and Models

Once we've developed immersive visions of desirable futures through personas, locations, and artifacts, we can take the prototyping process even further by modeling the systemic dynamics and experiential frameworks underlying those futures.

This level of prototyping involves constructing interactive conceptual models and simulations that articulate the economic, technological, social, and metaphysical paradigms operationalizing the reimagined realities. We develop experiential prototypes embodying the theories of change and ontological frameworks required to sustainably manifest the visionary futures.

There are many techniques for this level of systemic prototyping:

Future Systems: Modeling using computational modeling, simulations, and games, we can prototype the interrelated dynamics of alternative socio-technical, economic, and ecological systems aligned with the reimagined premises.

For example, we could develop an interactive simulation engine modeling how a future bioregenerative economy and resource distribution system could function, allowing people to experientially interface with its mechanics, feedback loops, and scaling principles. This makes the high-level vision tactile while stress-testing its viability.

CHAPTER 7   TOOLS AND TECHNIQUES FOR FUTURES THINKING EXPLORATION

Embodied Futures Practices: We could construct immersive frameworks allowing people to experientially inhabit the grounding metaphysics and epistemologies underlying the reimagined futures through embodied practices.

Perhaps we develop a "future body" social VR experience that guides participants through extended sensing protocols for recalibrating their phenomenological self-models and perceptual grammars to synchronize with a de-individuated mode of being. A way to get a felt sense of the ontological premises.

Speculative Societal Artifacts: On a more applied level, we could prototype artifacts simulating the logistical frameworks and cultural vernaculars that could emerge in a given reimagined future society.

For instance, we could develop a digital simulation of a regenerative circular resource provisioning system and peer governance protocols modeling how material/energy flows and citizen accountability could operate in a future world without money or centralized institutions.

Overall, the aim is to make the societal DNA of the reimagined futures as immersive and experientially resonant as the lifeways they seek to engender. We develop resonant metamodels and participatory frameworks embodying the integrative dynamics and epistemologies allowing the high-level possibilities to viably self-actualize.

This level of conceptual prototyping should be a fundamentally transdisciplinary practice drawing from diverse knowledge streams including

- Complexity sciences and living systems theory
- Embodied cognitive sciences and non-ordinary consciousness studies
- Theoretical physics and information sciences
- Biomimetic computing, regenerative design, systems ecology

- Indigenous cosmologies and integral metaphysics
- Social change praxis, futures studies, and prefigurative politics

Only by deep synthesis across these domains can we develop prototypes attuned to the generative codes and interweaving patterns required for radically alternative realities to harmonically emerge.

This work should also be highly collaborative and open source, inviting diverse human and nonhuman participants to iteratively experience, remix, and extend the metamodels. We construct immersive frameworks continually refined through cycles of participatory dialogue, phenomenological resonance testing, and regenerative revision.

When combined with the previous work of developing visions, personas, and artifacts, this multilayered prototyping allows us to navigate coherently between the most intimate and cosmic scopes of envisioned futures. We create cascading portals from speculative artifacts all the way up to metamodels of integrative dynamics embodying radically reimagined realities.

7.3 Summary

This chapter delved into the strategic and creative methodologies for developing and analyzing future scenarios, with a strong emphasis on tools like the Futures Wheel and Futures Cone. These techniques are crucial in mapping out potential future trajectories, identifying key uncertainties, and exploring the cascading impacts of various scenarios. The chapter also highlighted the importance of creating immersive and tangible representations of future possibilities through the development of personas, locations, and artifacts. By constructing detailed multimedia portraits and prototypes, futures thinking becomes more than just an intellectual exercise—it becomes a way to viscerally engage with and experience the potential realities that may emerge.

CHAPTER 7   TOOLS AND TECHNIQUES FOR FUTURES THINKING EXPLORATION

Moreover, the chapter introduces advanced techniques for prototyping future systems, emphasizing the need for transdisciplinary collaboration. This includes modeling socio-technical and ecological systems, creating speculative societal artifacts, and developing experiential frameworks that embody the underlying dynamics of reimagined futures. The goal is to create resonant, participatory models that bridge the gap between theoretical possibilities and lived experiences, enabling a more coherent and practical approach to shaping the futures we wish to create.

CHAPTER 8

Strategic Foresight and Backcasting

The immersive visions, artifacts, and metamodels we have prototyped are not just creative speculation—they represent possibilities we can consciously choose to make reality through intentional action. But actualizing the most inspiring and regenerative futures requires more than articulating the destinations. We need actionable frameworks for steering our trajectory in their direction from the present moment.

This is the domain of strategic foresight and backcasting—toolkits for translating our desired future visions into applied pathways and strategic decisions to start embodying them into existence. While the previous prototyping work was focused on expanding the possibility of space, these methods are about converting those possibilities into tangible, accountable praxis.

Strategic foresight practices allow organizations and change-makers to embed the long-view of participatory futures literacy into their strategic planning and innovation pipelines. By applying disciplined foresight, we can future-proof systems, products, and go-to-market strategies against paradigm shifts and blindspots in the present. It is how we align investments, priorities, and decision-making toward the biggest waves of change and transformation on the horizon.

CHAPTER 8   STRATEGIC FORESIGHT AND BACKCASTING

Backcasting takes this future-conscious orientation even further by reverse engineering pathways backward from envisioned endpoints. Rather than just strategizing based on extending present-day operating models, it allows us to map out the ramified steps, interventions, and paradigm shifts required to transition toward regenerative and equitable futures starting today. It's a way to get ahead of the curve by embodying the future into the present iteratively over time.

Together, strategic foresight and backcasting turn the immersive visions and artifacts we have prototyped from inspirational ideals into applied trajectories and accountable action portfolios. They convert the abstract into purposeful praxis linking the transcendent and immanent, the aspirational and contingent, into a participative feedback loop of shaping the futures we actually want to create.

In the coming sections, we will unfold specific frameworks and best practices for applying strategic foresight disciplines and backcasting pathways to industries, organizations, communities, and even personal praxis. We will explore techniques for

- Embedding futures literacy into brand/product innovation pipelines
- Generating Strategic Futures Playbooks to futureproof go-to-market strategies
- Mapping transition roadmaps and root causes to backcast from desired futures
- Prototyping actionable interventions to start embodying the future today

These methods turn the inspirational potentialities into living, evolving artifacts. They make the transcendent immanent through pragmatic frameworks, decision-heuristics, and portfolios of applied metamodernism continually refreshing our personal and collective praxis.

CHAPTER 8   STRATEGIC FORESIGHT AND BACKCASTING

8.1 Strategic Foresight Principles and Approaches

You probably might have also seen it over and over with companies that got blindsided by emerging technologies, new competitors, cultural shifts, you name it. They were too focused on short-term quarterlies and operations to perceive the big changes brewing on the periphery. That's why an intentional strategic foresight practice is so critical nowadays. It's about developing the organizational habit of consistently scanning the horizon for signals of change—those early ripples that could grow into disruptors of the entire industry in 5–10 years. Whether it's monitoring new scientific research, tracking fringe subcultures, or analyzing socio-political currents, foresight work is tuning into the present day periphery for those first glimpses of emerging futures.

As you might suspect already, we cannot just passively monitor, we need frameworks to make sense of the signal flows and derive insight from them. That's where methods like environmental scanning, trend analysis, and scenarios come in. For example, a renewable energy company might use comprehensive horizon scanning to map out a scenarios ecosystem. They might explore intersecting trajectories of climate impacts, policy shifts, new technologies, and more to pressure test their long-term strategic roadmaps against multiple future realities.

Modeling out those plausible futures can give them a comparative lens to evaluate their investment decisions—would this new biofuel program be viable in a fossil-stubborn scenario versus a rapid decarbonization one? You get the full 360° strategic picture when you can simulate multiple potential futures rather than just banking on one.

But scenarios alone are not enough. We also need approaches for adding that human, cultural texture. That is when participatory forecasting can be used to bring diverse stakeholders into the foresight process. This can be done by running workshops with futurists, engineers, activists,

CHAPTER 8   STRATEGIC FORESIGHT AND BACKCASTING

and community leaders to cocreate immersive artifacts and experiential scenarios together. Like prototyping what an off-grid coastal settlement could look and feel like, embedding those cultural nuances into the strategic foresight makes it way more actionable.

The best foresight work does not just analyze the future in abstract terms but it cultivates an organizational embodiment of emergence, where futures thinking gets baked into the cultural fabric. We are training perception to navigate complexity and nonlinear change as a core capacity. That is how we can avoid getting trapped by past patterns of success and keep evolving as the terrain keeps shifting.

One company that truly embodied strategic foresight reshaping their innovation trajectory was Microsoft. In the mid-2000s, they were heavily focused on iterating Windows and Office products on traditional computing platforms. However, their leadership recognized the need to explore how exponential technological forces and cultural shifts could reshape the entire computing landscape long-term.

They established an explicit futures research program called Microsoft Office Labs that deployed environmental scanning, cultural ethnographies, technology road-mapping, and speculative prototyping. Researchers explored intersecting trajectories in areas like ubiquitous computing, ambient intelligence, biotech integrations, and new human–machine interfaces.

From these futures work emerged product concepts and experience prototypes that seemed wildly audacious at the time, like the Microsoft HoloLens augmented reality system and reactive user interfaces driven by data streams. But these became the seeds for Microsoft's strategic pivot into pioneering mixed reality, voice interfaces, and more human-centered intelligent cloud platforms.

Their innovation pipeline transformed from just incrementally improving Office to prototyping entirely new computing paradigms attuned to broader socio-technological shifts surfaced by their strategic

CHAPTER 8   STRATEGIC FORESIGHT AND BACKCASTING

foresight work. The foresight research legitimized long-term R&D investments that future-proofed their product roadmaps for decades rather than just the next device cycle.

Microsoft exemplified how strategic foresight can reveal emerging multidisciplinary forces and human dynamics that shape technological needs. By committing serious resources to prototyping products for those longer-term futures, they could continually evolve rather than get blindsided by disruptions to their core business models.

We need to remember through that the real power of strategic foresight emerges from the inner practice—staying alert to emerging potentials, even in the faintest present-day signals, and embodying those capacities throughout an organization. When that clicks, companies quit reacting to futures that happen to them and start creating the futures they're meant to exemplify.

8.1.1 Application in Market and Technology Trends

One of the core domains where strategic foresight provides immense value is surfacing the technological and market discontinuities that could disrupt an organization's future operating environments. Executives often get blindsided by paradigm shifting innovations or tidal waves of changing consumer behaviors because they are too focused on incremental thinking and quarterly performance. Strategic foresight helps expand the aperture.

A key foresight practice in this field is environmental scanning and horizon scanning, systematically monitoring a wide range of signals for emerging issues and discontinuous forces from the periphery. This could involve tracking scientific research, mapping subculture dynamics, analyzing policy developments, or studying fringe social currents. The key is tuning into those faint ripples in the present that could grow into massive waves of change and impact in the future.

CHAPTER 8   STRATEGIC FORESIGHT AND BACKCASTING

For example, consumer goods giant Procter & Gamble uses environmental scanning approaches to explore how changing values, new technologies, and societal shifts could reshape consumer behaviors and market categories long-term. Their Future Waves thought-leadership hub analyzes emerging undercurrents in everything from wellness movements to household dynamics to map how their brands may need to evolve.

Another strategic foresight method P&G employs is trend impact analysis—quantitatively modeling out how current technological, economic, and cultural forces could impact their product lines and supply chains in upcoming decades. They can simulate different scenarios like distributed additive manufacturing converging with circular economies to evaluate fundamental disruptions to their manufacturing and go-to-market strategies.

In technology domains, corporations like Samsung deploy techniques like technology sequence analysis and data mining to perceive potential trajectories of scientific breakthroughs. They scour research publications, patent filings, and other data sources to construct probabilistic models of how fields like AI, nanotech, or biocomputing could progress and enable next-generation innovations. This guides their long-term R&D investments and product roadmaps.

Larger tech firms like Google also leverage strategic foresight through dedicated teams running continuous scanning, trend analysis, and scenario modeling. For example, their AI research group explores potential futures stemming from intersecting trajectories in fields like neurotechnology, data/sensor proliferation, and distributed computing to continuously reimagine future AI capabilities and applications. Their strategic foresight work imagines paradigm shifts like ambient intelligence environments and neural interfaces decades before commercial manifestation to guide moonshot R&D programs.

Scenarios and dynamic simulations are critical to strategic foresight for evaluating how disparate forces could converge into new future realities. Companies like Ford and Volkswagen run immersive scenario exercises to prototype how autonomous mobility, new human-machine interfaces, renewable energy transitions, and shifting urban environments could completely reshape the future of transportation and vehicle experiences over the next 20-30 years. This expansive, possibility modeling approach radically expands their strategic contexts beyond just the near-term vehicle product cycles.

The key is using strategic foresight methods in an integrated, futures-conscious way—continuously scanning for emerging signals and trends, then pressure-testing strategies, products, and business models against multiple scenarios modeling how those forces could reshape consumer realities. It's an organizational capacity for expanding perception to envision and choreograph strategies across potential future discontinuities, rather than just reacting to them as disruptions.

8.1.2 Creating a Strategic Futures Playbook for Brands and Products

While techniques like environmental scanning and trend modeling help organizations perceive the signposts of change, truly future-proofing brands and product portfolios requires more immersive, participatory strategic foresight work. This is where companies develop comprehensive "futures playbooks" that combine scenarios, artifacts, systems maps, and diverse expert contributions into holistic foresight territories.

A core practice in this vein is participatory forecasting utilizing methods like immersive workshops, design fiction prototyping, and multimedia artifacts. Rather than just analytical scenarios on paper, these approaches bring the visions to life in visceral, experiential ways through creative storytelling and speculation.

For example, global brand strategy firm Idea Couture runs foresight workshops for companies like Samsung, Adidas, and Nestlé. Their Kaleidoscopic Worldbuilding process immerses participants in futures

CHAPTER 8   STRATEGIC FORESIGHT AND BACKCASTING

storytelling and artifact prototyping to add resonant cultural texture to scenarios. Instead of just abstract scenarios, clients get to inhabit storied futures through sensory narratives, product concepts, and multimedia simulations customized to their innovation domains.

Similarly, firms like TillSpace Partners leverage practices like causal layered analysis and systems modeling to map out the deeper paradigm shifts and systemic forces emerging as the contexts for brand and product futures. Their Foresight approach helps clients like Audible, Spotify, and Amazon unpack the larger metamodernist currents rippling through areas like the creator economy, voice/smart environments, and the passion economy. These metamaps are brought to life through explorable digital ecosystem models that creatively stress-test brands against holistic future shifts.

Similarly, strategic foresight firm Wikistrat runs crowdsourced sessions that gamify predictive futures intelligence through cloud-based simulations. These massive multiplayer online scenarios allow diverse experts and stakeholders across the world to collaboratively model how emerging signals could converge into new future realities, providing probabilistic data to inform brand strategies.

Combining methods like these, strategic foresight teams craft expansive playbooks that combine hard data analysis with immersive multimedia artifacts walking brands through multiple alternative future contexts. These integrative playbooks holistically map out the experience Journey of evolving consumer needs, behavior shifts, technology implications, systemic disruptions, and metamodernist currents that could reshape a company's product ecosystem in the upcoming decades.

The foresight work does not stop at scenarios. It amounts to prototyped future worlds that brand teams can inhabit and pressure test their pipeline strategies against from diverse firsthand perspectives. Automakers get to experientially role-play in speculative driving simulations shaped by post-urban dynamics. Fashion houses envision new apparel languages emerging from nano-material science, avatars, and human/AI convergence patterns. What could have been unseen disruptions become creatively choreographed opportunity spaces.

Investing in these pluralistic, viscerally resonant futures playbooks prepares organizations to get ahead of the massive changes redefining the contexts for brands and products over long-time horizons. Rather than rigidly chasing quarterly roadmaps, they cultivate an adaptable futures consciousness that continually reorients innovation pipelines as the terrain evolves in novel directions.

8.2 Backcasting: Charting the Path to Desired Futures

All the foresight work in the world analyzing potential futures is only as valuable as our ability to shape reality toward the ones we actually want to create. This is where the practice of backcasting becomes essential.

While forecasting explores what future outcomes seem probable based on current trajectories, backcasting takes a fundamentally different approach—envisioning desired future endpoints and then reverse-engineering the transformations required to manifest them from the present. It's a way to construct actionable pathways to bring our most inspiring and ethically aligned visions of the future into reality.

The core backcasting philosophy is simple yet profound: do not just predict the default "future" trajectory, choose the one you prefer, then chart out the ramified policies, investments, innovations, and behavior changes that would be required to actualize that future scenario over time. It's a solution-oriented methodology, starting with the solutions you want and working backward to strategize their implementation.

This reverses the typical institutional mindset stuck in the inertia of path dependencies and incremental thinking. Rather than just extending trendlines from the current operating models, backcasting encourages

CHAPTER 8 STRATEGIC FORESIGHT AND BACKCASTING

audacious imagination, dreaming into the worlds that could be forged through intentional system redesigns and metamodernist interventions over longer timescales.

There is a wide array of specific methodologies that can be employed in backcasting processes, which we will explore in the coming sections. But at its core, backcasting represents a fundamental reframing, rather than surrendering agency over our collective futures to the path of least resistance, we become conscious choreographers and designers shaping the world into progressively preferred realities through intentional choice and effort over time.

In essence, backcasting is how we translate our inspirational visions and transformative scenarios into applied metamodernism, embodying the transcendent into the immanent, the imaginal into the pragmatic, in grounded, phased, and accountable ways. It is the technology for converting abstract ideals into lived experience through diligent, multigenerational trajectory setting.

In the next sections, we will dive into some of the key frameworks and methodologies that comprise skillful backcasting—from road mapping and root cause analysis to developing implementation steps that turn ideation into sustained praxis.

8.2.1 Understanding and Applying the Backcasting Methodology

At its core, the backcasting approach involves a fairly straightforward conceptual framing: define a future vision or normative scenario representing the desired endpoint, then work backward to determine what sequences of transformations would allow that future to be realized over time.

The process typically begins by developing rich, multimedia visions or artifacts that vividly portray the preferred future state across different dimensions—whether that's sustainable human settlements, revitalized

CHAPTER 8   STRATEGIC FORESIGHT AND BACKCASTING

ecosystems, technological paradigm shifts, or entirely reimagined political/economic frameworks. These artifact worlds don't just describe the end vision, but saturate participants in the qualitative textures and experiential realities they would inhabit.

From these immersive visions, backcasting teams then begin mapping out sequences of technological, policy, investment, and societal/behavioral transformations that would progressively bridge reality toward that preferred scenario over future time horizons—typically looking 20-50 years into the future and working backward.

For example, if the vision is proactively transitioning a city into a decarbonized, regenerative urban ecology by 2070, the backcasting process might identify transitional roadmaps like

- Large-scale green infrastructure overhauls between 2025 and 2035
- Renewable microgrids and energy democratization from 2030 to 2045
- Rewilding and ecological restoration corridors from 2035 to 2055
- Public mobility/housing remapping between 2040 and 2060
- Circular economic and social policy transformations from 2045 onward

With each phase, subject matter experts deeper unpack the technical, legislative, and socio-economic changes required, continuously iterating the roadmap into increasingly granular and actionable detail. Interdependencies and stakeholder dynamics are mapped through causal analysis and systems modeling.

CHAPTER 8 STRATEGIC FORESIGHT AND BACKCASTING

A key part of backcasting is conducting root cause analysis on current systemic behaviors, lock-ins, and path dependencies that could act as impedances or barriers to realizing the vision if left unaddressed. This surfaces key leverage points that need to be fundamentally transformed, evolving beyond surface-level symptom responses.

For the urban ecology vision, root causes like economic extractivism, siloed governance, and societal energy/material dominance behaviors would need to be deconstructed and redesigned through interventions years or decades earlier. Backcasting reveals these deep, transdisciplinary changes and system acupuncture points.

Throughout this reverse-engineering, backcasting develops portfolios of interventions addressing different root causes and trajectory transformations. Within these are tailored policy packages, technology roadmaps, financing strategies, public education/media interventions and other transdisciplinary solutions staged over time. It's essentially mapping an evolving, multidimensional "solution ecology" for materializing the vision.

Crucially, backcasting is not just analyzing the desired future in isolation. An iterative part of the process is pressure-testing the roadmaps against other potential future scenarios and disruptions using strategic foresight modeling. This ensures the transition pathways remain resilient and dynamically evolving even as unexpected changes emerge.

When done diligently, the output of a backcasting process is an evolutionary transformation pathways portfolio—a multidimensional, living roadmap of interwoven policies, technological solutions, cultural narratives, investments, and interventions continually refined to bridge reality toward the manifestation of the preferred vision over future timeframes.

While ambitious, the backcasting process is firmly rooted in pragmatic, evidence-based strategizing. By working backward from endpoint visions, it ensures long-view alignment where short-term projects and reforms

CHAPTER 8   STRATEGIC FORESIGHT AND BACKCASTING

accumulate into cohesive trajectories for realizing transformed future systems. It's the intentionality and trans-disciplinary rigor for actualizing our most inspirational dreams into reality over time.

8.2.2 Roadmapping and Root Cause Analysis

Two of the core techniques employed in comprehensive backcasting efforts are roadmapping and root cause analysis. These methods help map out the intricate transition pathways and systemic interventions required to progressively evolve from the current reality toward realizing the envisioned preferred future state.

Roadmapping involves choreographing the sequenced rollout and implementation of different policies, technological deployments, financial instruments, societal shifts, and other transformative solutions over future time periods. It is essentially scheduling and mapping the "solution ecology" that gets staged and phased to reshape systems incrementally.

For example, if the vision being backcast is establishing sovereign resilient bioregions with decentralized distributed economies by 2070, the roadmap might schedule phases like these:

2025–2035: Localized food/water infrastructure, community trusts, transition towns

2030–2040: Municipal renewable mesh grids, mobility/work transformations

2035–2045: Regenerative production centers, public banking, edu-cultural shifts

2040–2060: Bioregional self-governance models, new economic paradigms

Within each phase, the roadmap unpacks specific policies, technological deployments, skill transformations, civic/social innovations, and other progress metrics that need to be rolled out incrementally to build toward the next transition stage. Interdependencies between strands are mapped and interventions co-scheduled.

CHAPTER 8   STRATEGIC FORESIGHT AND BACKCASTING

Roadmapping helps delineate actionable steps while visualizing the intentional, evolving sequences required to reshape entrenched systems into radically new future forms over time in pragmatic, resilient ways. It creates a cohesive, multidimensional master plan continually evaluated against changing landscape conditions.

Complementing the roadmapping process is root cause analysis. This involves excavating the deeper systemic drivers, entrenched behaviors, and path dependencies currently preventing or creating impedances toward realizing the envisioned future reality if left unaddressed. It's about getting to the source roots rather than reacting to surface-level symptoms.

For the bioregional future, root cause analysis might identify sources like these:

- Carbon-extractive economic principles encouraging centralization
- Societal energy/material dominance behaviors and perceptions
- Restrictive legal/policy frameworks around food, finance, education, etc.
- Cultural narratives and values promoting excessive consumption
- Community disintegration from hyper-individualization

Rather than just treating symptoms, the root cause analysis reveals these interrelated systemic sources fueling and locking in the current undesirable trajectories as levers that need to be profoundly transformed. Each root then becomes a portfolio of transformative interventions strategically scheduled and mapped across the transition roadmap.

For example, societal energy dominance behaviors might be intervened upon through public education campaigns, community skilling efforts, new infrastructure incentives, and even childhood development narratives, all choreographed over phased timelines. Policy reforms, financing initiatives, cultural artifacts, and more all become sequenced to reshape that systemic root.

Combining roadmapping with root cause analysis, the backcasting process holistically maps not just hopeful ideals, but the arduous, transdisciplinary pathways for evolutionary systemic metamorphosis. It reveals the transformative actions required to reshape entrenched realities into sustainable, ethically grounded futures over generational time frames in proactive yet pragmatic progressions.

The roadmaps become both a general project plan for coordinating cross-sector efforts and solution portfolios as well as comprehensive "unlearning" pathways for transitioning societies out of degenerative systemic behaviors into revitalized, coherent future models and consciousness states.

8.2.3 Backcasting Actionable Roadmaps from Preferred Futures

The ultimate aim of any rigorous backcasting process is to translate the inspirational facets of our preferred future visions into tangible, phased roadmaps for actualization within our current realities and cognitive/resource landscapes.

This involves taking the generalized transition sequences and root cause portfolios mapped in the previous backcasting stages and rendering them into explicitly defined, bounded action projects that can be implemented by specific entities over scheduled timelines. We are essentially breaking down the macro-level roadmap into prioritized sprints and intervention milestones.

CHAPTER 8   STRATEGIC FORESIGHT AND BACKCASTING

For envisioned futures like localized circular economies, regenerative urban environments, or shifts toward decentralized autonomous organizations, backcasting teams derive actionable pathways that reflect

Short-Term Catalysts (1–3 Years)

- Pilot projects to prototype core concepts
- Public education and narrative campaigns
- Initial legal/policy reforms and incentives
- Financing vehicles and funding pools

Medium-Term Systemics (5–10 Years)

- Integrated tech/infrastructure deployments
- Public institutional redesigns/reshapings
- Economic incentive/regulatory overhauls
- Social/cultural acupunctural efforts

Long-Term Evolutionary Paradigms (10–25 Years)

- Comprehensive post-transition governance models
- Evolved societal consciousness/skill embodiments
- Regenerative production/provisioning systems
- New scientific/innovation paradigms and R&D programs

The key is deriving explicated solutions, milestones, roles, and orchestrated pathways from the macro-roadmaps, clearly defined and scheduled actions that accumulate into realizing the visionary future systems over time. Budget projections, entity owners, measurement frameworks, and contingencies all get mapped into actionable implementation cycles.

For example, transitioning toward decentralized autonomous organizations might involve backcasting milestones like

Short-Term (1–3 Years)

- Regulatory reforms for distributed entrepreneurship
- Collaboration/meritocracy skill training programs
- Open source DAO governance prototypes/sandboxes

Medium-Term (5–10 years)

- Tax/finance transformations for localized stakeholding
- Transition of core public services into DAO equivalents
- New economic/incentive frameworks restructuring labor
- Public blockchain identity/reputation systems

Long-Term (10–25 years)

- Comprehensive DAO cities and bioregions
- Distributed production/provisioning networks
- Social reputation/meritocracy models for education/governance
- Embodied cultural shifts around individualism/collectivism

By rendering backcasting roadmaps into explicitly scheduled and defined initiatives, policies, infrastructure overhauls, institutional redesigns, and interwoven interventions, we create a multi-stakeholder implementation gameplan for actually making the visionary futures manifest from the present day.

The backcasting process does not end with the roadmap delivery. An equally critical phase is establishing governance, measurement, and accountability frameworks to track and evaluate progress against the scheduled transformation pathways as they get rolled out.

This typically involves technology roadmapping initiatives to coordinate innovation pipelines, policy tracking mechanisms, continuously updated public education/narrative campaigns, multisector financing and resource coordination, and other mechanisms to ensure society-wide orchestration and embodiment of the interventions across different disciplines.

Progress measurement also circles back to inform roadmap iterations. As realities evolve, the backcasting roadmaps and initiatives need to be retested, updated, and rolled out in new tranches to simplify and navigate transition obstacles and emergent change dynamics.

Backcasting empowers societies to become the authors of transcendent, ethically sculpted trajectories.

8.3 Summary

This chapter explored the powerful methodologies of strategic foresight and backcasting, which serve as essential tools for transforming visionary futures into actionable realities. Strategic foresight allows organizations to anticipate emerging trends and paradigm shifts, ensuring that their strategies are resilient and aligned with long-term goals. By continuously scanning the horizon for signals of change and using participatory approaches to incorporate diverse perspectives, organizations can develop immersive, forward-looking strategies that are better equipped to navigate the complexities of an uncertain future.

CHAPTER 8   STRATEGIC FORESIGHT AND BACKCASTING

Backcasting complements foresight by working backward from desired future scenarios to identify the necessary steps and interventions required to achieve those outcomes. This process involves creating detailed road maps that outline the sequence of actions needed to transition from the present to the envisioned future. By integrating root cause analysis, roadmapping, and strategic implementation frameworks, backcasting provides a practical, systematic approach to realizing transformative futures. Together, these methods enable organizations and communities to proactively shape their trajectories, translating aspirational visions into sustainable, actionable plans.

CHAPTER 9

Synthesizing Futures Practice

This chapter offers a worked, instructional example to show how scanning ➤ sensemaking ➤ scenarios ➤ artefacts ➤ backcasting ➤ decisions fit together. It is not a formal study; in practice, teams should begin with evidence gathering and stakeholder engagement, then use the structures here to integrate their own research into decisions.

So far, we have explored a wide array of methods and frameworks for developing futures literacy from scenario modeling and prototyping to strategic foresight and backcasting. Each technique has equipped us with vital tools for perceiving potential trajectories, creating resonant visions, and mapping pathways to realize the futures we actually want to exist in.

But as with any rigorous practice, the true mastery emerges when we can synthesize and apply these methods in coordinated, contextual ways. This chapter represents that integration, guiding you through an applied comprehensive futures project where you will blend forecasting, backcasting, strategic planning, ethical frameworks, and more into a cohesive body of futures work.

We will start by exploring how futures design must be deeply rooted and contextualized within the specific domains, communities, and cultural narratives it's aiming to impact. The same generalizable methods take on nuanced applications when applied to urban planning versus consumer

CHAPTER 9   SYNTHESIZING FUTURES PRACTICE

product futures versus regenerative agriculture. Authentic futures practice demands embodying the voices, values, and lived realities of the people and places you're envisioning futures for.

From that grounding of context and "place," we will then select a concrete focal area, problem space, or opportunity domain to concentrate your comprehensive futures project within. This could range from future models for distributed education or smart cities to economic rebundling or democratic social media—any complex terrain ripe for applying synthesized futures methods.

With that focal area framed, you will work through building multimedia scenario artifacts, developing strategic road-mapping and visioning, pressure-testing strategies through backcasting and foresight modeling, and deriving pragmatic portfolios of actionable interventions to start actively transitioning toward your envisaged future models today.

The team will integrate methodologies, contextualize visions based in specific places/communities/cultures, navigate ethical tensions, and seed actionable pathways to start manifesting inspirational futures models into reality over time.

The goal is for you to emerge with a comprehensive futures project robustly addressing the multifaceted human, technological, and planetary dynamics that will shape the terrain of your chosen focal domain. One that balances speculative imagination with grounded, contextual praxis. One that reflects your unique voice as an emergent leader and world-builder while drawing from a rich toolkit of futures methods.

Scope and Evidence Notes:

This section presents an instructional example that demonstrates how to apply foresight methods end-to-end (scanning ➤ scenarios ➤ artifacts ➤ backcasting). It is not a formal research study. To sound the example, the team seeded it with public data and reports (See Evidence Base below) and kept a traceable chain from signals to implications. Use the structure, then replace the placeholders with your own research when running this in practice.

CHAPTER 9   SYNTHESIZING FUTURES PRACTICE

Evidence Base (Illustrative Sources Used to Seed the Example):

- Global EV adoption, policy, and battery trends: IEA Global EV Outlook; BloombergNEF Electric Vehicle Outlook.

- Supply-chain and materials risk: IEA/USGS battery minerals; OECD trade briefs.

- Grid and charging infrastructure: IEA electricity reports; national regulator stats.

- Climate scenarios framing: IPCC AR6 synthesis.

Method Transparency:

- Scanning organized with STEEP+V/PESTLE; signals logged with source, date, and relevance.

- Scenarios built via axes of uncertainty and elaborated with Futures Wheel.

- Backcasting produced options, triggers, and staged experiments.

- Decisions recorded with KPIs and review cadence.

Limitations and Next Steps:

This example comprises analysis and omits stakeholder research. A real project would add primary research (interviews, surveys, expert elicitations), test assumptions with prototypes/pilots, and document distributional impacts (equity, more-than-human).

CHAPTER 9   SYNTHESIZING FUTURES PRACTICE

9.1 Contextualizing Futures in Place, Culture, and Community

Before we can apply futures design methods to any focal domain, we must first ground that work within the specific cultural narratives, community realities, and place-based contexts it aims to impact. Authentic futures practice demands embodying the voices, values, and lived experiences of the people and environments we're envisioning futures for.

This localized contextualization is vital because the human, geographic, and societal terrains futures emerge from are deeply rooted in the rich diversities of culture, history, and place-based identities around the world. The aspirations, beliefs, and circumstances shaping how communities relate to their potential futures are wholly unique depending on where you are.

For example, envisioning smart city futures for urban hubs in Europe versus Africa versus the Pacific Islands would require radically different contextualization. Each sphere represents distinct cultural metaphors, narratives of development/progress, and lived environmental realities that any smart city futures need to be attuned to for relevance.

Similarly, futures work for indigenous communities would necessitate situating that within their unique cosmologies, values around human/nature balance, and place-based wisdom inscribed over generations. The same generalizable futures methods take on nuanced applications when grounding those insights in specific community contexts.

This place-based attunement starts by deeply listening and developing resonant understandings of the cultures, identities, and circumstances communities are emerging from. It includes

- Immersing in local narratives, arts, histories
- Respecting indigenous epistemologies and ancestral wisdom

CHAPTER 9   SYNTHESIZING FUTURES PRACTICE

- Embodying lived experiences of geography and environment
- Attuning to customs, beliefs, and community value systems
- Understanding legacies of societal traumas and injustices
- Embracing evolving cultural metaphors around progress/change

Only once we have absorbed these rooted contextual elements can we begin translating our generalizable futures design methods in ways that add value and relevance to communities.

For example, we may draw inspiration from ancient regenerative practices while integrating frontier technologies and scientific wisdom. We could merge indigenous storytelling with multimedia scenario modeling. Immersive participatory prototyping might help us envision urban evolution accommodating both hyper-tech and traditional place-based knowledge synapses.

The key is ensuring that as futurists, we are not imposing extractive, colonizing mindsets onto our futures work. We must approach contexts with humility embedding our methods within the place-based intelligence to uplift endogenous visions of thriving human potential already present.

This grounds our futures practice in an ethos of cultural significance and empowerment. We are not parachuting ideas in, but facilitating communities in perceiving and activating the most transcendent possibilities already encoded in their identities, narratives, and environments. Our role becomes midwifing these into contextualized actions manifesting flourishing futures on their terms.

CHAPTER 9   SYNTHESIZING FUTURES PRACTICE

In this way, localizing our work in the richness of places, cultures, and communities becomes an essential first step before any techniques are applied. It ensures that regardless of the focal domain we select, our futures work emerges as a respectful, empowering praxis honoring the humanity of the people we aim to serve.

9.1.1 Role of Place in Futures Design

When we talk about designing the future, it's easy to get caught up in the grand narratives of technological disruption, societal shifts, or global trends. While these are undeniably important, I have come to believe that a critical piece of the puzzle is often overlooked: the role of "place."

By place, I don't simply mean a geographic location. Rather, it encompasses the rich context of physical environments, social structures, cultural norms, historical context, and the unique challenges and aspirations of the people who inhabit it. In essence, place is where the rubber of our grand visions meets the road of lived experience.

Authentic futures work does not abstractly parachute in decontextualized universal models, but roots its visions from the voices, values, lived experiences, and holistic worldviews of the specific peoples and places it ultimately aims to create better trajectories and possibilities for.

This could involve
Immersing in Narratives of Place

- *Interviewing elders and culture-keepers*
- *Studying oral histories and cosmologies*
- *Interfacing with land/nature to absorb lessons*

Embodying Vernacular Values and Wisdom

- *Decolonizing default mindsets and assumptions*
- *Embracing pluralistic ways of knowing and being*
- *Regrounding in locally attuned ethics and priorities*

CHAPTER 9   SYNTHESIZING FUTURES PRACTICE

Building from Participatory Visioning

- *Cocreating through community storytelling circles*
- *Collaborative prototyping using vernacular design*
- *Continual feedback loops between vision and place*

The aim is to cultivate deep empathic resonance between the envisioned futures and the textured place-based realities they're being creatively intervened within. It's about embodying holistic perspectivism—not just generating globally scalable scenarios abstractly, but rooting trajectories from the pluripotent soil of diverse cultural narratives and worlds.

For example, if aiming to envision future models of distributed education and youth empowerment, a futures designer would spend time

- *In deep storytelling practice with teachers from diverse cultures*
- *Honoring place-based intergenerational knowledge practices*
- *Understanding vernacular community dynamics and barriers*
- *Building prototypes with youth leaders and community guides*
- *Facilitating evolving vision/feedback loops for cultural resonance*

The resulting future visions and roadmaps wouldn't be decontextualized, Westernized models imposed as universal solutions. They'd be living narratives flowering from the unique soils, voices, and priorities of those specific places and peoples—harmonizing global futures principles with pluralistic contextual textures.

This is not always easy, especially given many Eurocentric modern defaults around individualism, objectivism, and one-world premises. It requires deep personal unlearning and ego dissolution into polycultural community attunement from the futures designer. But it's an indispensable part of the praxis.

9.1.2 Tools and Frameworks for Capturing Insights and Assessing Impacts

When situating our futures practice within the realities of actual places, communities, and cultural contexts, we need robust toolkits for empathetically capturing relevant insights and assessing potential impacts. Just as we can't develop products without user research, we can't ethically prototype futures without deeply understanding the specific social-cultural-economic fabrics we are aiming to transform.

A foundational set of methodologies for this crucial context-capturing revolves around participatory ethnographic approaches. The following are some examples of such frameworks:

Immersive Observation
Embedding ourselves within the communities and experiencing their day-to-day realities, routines, and vernacular activities firsthand through participant-observer research.

Narrative and Storytelling Studies
Analyzing the tales, idioms, mythologies, and other symbolic meaning-making cosmologies circulating as anchoring narratives within the social fabric.

Community Co-design
Collaborative workshops and human-centered design sprints done with diverse community members to empathize with their needs and dreams from the ground up.

CHAPTER 9   SYNTHESIZING FUTURES PRACTICE

By combining these qualitative, lived-experience methodologies, we develop grounded understandings of the rich cultural reference points, shared histories, and aspirational values inscribed into specific contexts. We cultivate embodied futures literacy about the people and places we're envisioning transformations for.

Quantitative data and analytics are also vital to round out these human contexts into holistic pictures. Community-scale surveys, systems modeling, and data-mining the digital trace artifacts can elucidate material conditions, socioeconomic patterns, and other mapping further cultural nuances. Geospatial sensing and mapping tools illuminate insights into the built environment, flows, and infrastructures.

Once we have this immersive 360-degree context-mapping, we can then develop robust frameworks for assessing and monitoring the multilayered impacts our futures work could have if realized within that environment:

Impact Narratives and Prototyping
Developing multimedia prototypes and experiential storytelling artifacts to viscerally simulate how new futures could manifest and ripple through specific contexts.

Ethical Scenario Modeling
Exploring fictional artifacts, personas, and worlds under different transformation scenarios to pressure-test for second-/third-order sociocultural implications.

Public Narrative and Discourse Analysis
Monitoring media landscapes and symbolic frames circulating to assess emerging adoption of preferred cultural narratives and mindsets your futures aim to instill.

Essentially, we're creating comprehensive, multidimensional sensing frameworks that use both quantifiable data and ethnographic insight to track vectors of impact across sociocultural, spatial, political, economic, and metaphysical planes.

This 360-degree context-mapping and impact assessment isn't just about checking ethical boxes. It's about developing nuanced futures empathy—embodying the diversity of peoples' lived experiences so we can sculpt visions with relevance without appropriating or extracting. It ensures our transcendent imagination syncs up with immanent realities of place.

As with any craft, mastery comes through embracing a beginner's mindset and humility. We are not infallible wizards, but diligent world-renewing students and cocreators aiming to support thriving futures in symbiosis with local and planetary contexts, not in opposition to them.

9.2 Final Project: Comprehensive Futures Project

For your culminating futures practice application, you will develop a comprehensive futures project that synthesizes multiple methods and frameworks into an integrative body of work. This project will showcase your ability to blend techniques like scenario modeling, multimedia prototyping, strategic foresight, and backcasting in a coordinated, contextual way.

The comprehensive project will entail the following.

9.2.1 Selecting a Focal Issue/Domain

You will choose a specific issue, opportunity area, or complex problem domain to concentrate your futures work within. This could range from future models for education or healthcare to economic restructuring, urban resilience planning, technology ethics, or reimagining governance frameworks. The key is selecting a rich terrain with multifaceted human, technological, environmental, and societal implications to fully flex your integrative futures muscles.

9.2.2 Applying Futures Methodologies

Prerequisite: Evidence and Engagement First

Begin with research—secondary sources (official stats, industry reports, academic work) and primary inputs (stakeholder interviews, workshops, expert elicitation). Log each item with source/date and a one-line relevance note.

1. Environmental Scanning (STEEP+V/PESTLE)

    Inputs: Curated sources, horizon-scan feeds.

    Activities: Classify signals under Social, Technological, Economic, Environmental, Political +Values (or Legal if using PESTLE); note weak vs. maturing signals.

    Outputs: Signal log, emerging patterns/themes.

2. Stakeholder and Domain Research

    Inputs: Interview guides, participatory sessions, desk research.

    Activities: Map stakeholders, motivations, power and equity concerns; document lived experience insights.

    Outputs: Stakeholder map; evidence-backed insights and tensions.

3. Sensemaking ➤ drivers and uncertainties

    Inputs: Signals + stakeholder insights

    Activities: Cluster patterns; drive drivers of change; identify critical uncertainties; use tools like Futures Wheel to trace consequences

CHAPTER 9   SYNTHESIZING FUTURES PRACTICE

Outputs: Short list of drivers; 2–3 uncertainty axes; implications list.

4. Scenario Building

   Inputs: Drivers+uncertainty axes

   Activities: Construct 3–4 plausible scenarios; write narrative briefs; pair each with early indicators

   Outputs: Scenario set, indicator list, comparison table.

5. Prototyping (Experiential/Artifactual)

   Inputs: Scenario narratives

   Activities: Create artifacts, story worlds, or simulations that make implications tangible; test comprehension with nonexperts.

   Outputs: Artifacts and learning notes tied to each scenario

6. Backcasting and Decision Pathways

   Inputs: Preferred Criteria+Scenario Implications

   Activities: Define options, triggers, and no-regret moves; sketch transition pathways from the future back to today.

   Outputs: Option set with trigger thresholds, near-term experiments, and risk/impact notes.

7. Governance and Review

   Inputs: KPIs and signal dashboard

CHAPTER 9   SYNTHESIZING FUTURES PRACTICE

Activities: Set review cadence (quarterly/semiannual) to refresh scans, update scenarios, and adjust options.

Outputs: Operating rhythm; roles and owners.

Tip   *Keep traceability visible throughout: Source ➤ Insight ➤ Implication ➤ Option/Trigger.*

9.2.3 Presenting and Peer Reviewing

The final component is presenting and peer reviewing your comprehensive futures projects. You will have the opportunity to share your visions, prototypes, strategic plans, and backcasting roadmaps—opening them up to questioning, feedback, and collaborative refinement.

This review process will surface potential blind spots, integrate additional perspectives, and allow benchmarking your work against other futures projects across domains. The goal is to pressure-test the integrative rigor and ethical grounding of your comprehensive approach.

By the end, you will have an enriched, foundational grasp of the continual practice of applying futures methods in contextual, multifaceted ways.

9.2.4 Selection of a Focal Area for Futures Design

The first step in constructing a comprehensive futures project is determining what specific issue, opportunity space, or socio-technological terrain you want to deeply explore and design potential futures around. This focal area will essentially become the canvas that all your integrated methodologies get applied upon.

CHAPTER 9   SYNTHESIZING FUTURES PRACTICE

The options are quite expansive, but some example domains others have concentrated futures projects within include

New Economies/Industrial Transformations

- *Decentralized/distributed economic models*
- *Transition paths beyond extraction capitalism*
- *Future production/provisioning ecosystems*
- *Evolution of markets, exchange, and incentives*

Technological Paradigm Shifts

- *Futures of AI, human–machine symbiosis*
- *Next-gen computing/networking paradigms*
- *Biotechnology and ecological regeneration*
- *Breakthrough energy and climate solutions*

Social Systems and Human Development

- *Equitable education and knowledge economies*
- *Decentralized governance and democracy*
- *Evolution of cities, built environments, mobility*
- *Public health, mind/body, community vitality*

You will want to select a focal area that has sufficient complexity, with intersecting uncertainties around technological forces, policy impacts, human dynamics, environmental tensions, and other interwoven dynamics. Oversimplified issues tend to result in threadbare scenarios and visions.

At the same time, you will need to scope the area reasonably so it's bounded enough to apply all the different futures methodologies in a cohesive, contextually relevant way within the project timeframe. Trying to model the entire future of human civilization may be a bit too expansive for this purpose.

CHAPTER 9   SYNTHESIZING FUTURES PRACTICE

A good litmus test is identifying a focal area that sits at an intriguing systemic intersection requiring reimaginative solutions—like the future of food at the nexus of agriculture, climate, public health, supply chains, and consumer behaviors. That frames a rich complexity to dive into creatively without being abstract or unmanageably vast.

Another consideration is selecting a focal area you have some familiarity with or intellectual passion for. It's much easier to immerse yourself in the deep uncertainties and situate viable visions if you already have a contextual foothold in that domain to build deeper from.

For example, if you have educational experience and care about accessible knowledge, focusing on "Futures of Learning Ecosystems" could be very fertile territory to imagine new paradigms and solutions for. If you're passionate about cities and mobility, homing in on "Urbanomic Renaissance" as the focal area frames an evocative sandbox.

The goal is to scope a context that hooks your authentic curiosity and care, while still maintaining sufficient complexity and systemic richness worthy of a robust, multifaceted futures design process.

Once you have selected that focal issue or domain, the rest of this chapter will guide you through applying various futures methods to actively model out potential scenarios, develop strategic pathways, prototype solutions and artifacts, and construct comprehensive visions and transition roadmaps addressing that terrain in tangible, actionable ways.

So don't limit yourself, but also be reasonably scoped. This is likely one of the first applied futures projects you'll develop rather than a thesis, so aim for a fertile context you can immerse and expand yourself within for this initial foray, while still leaving ample creative territory to explore further down the line.

CHAPTER 9   SYNTHESIZING FUTURES PRACTICE

9.2.5 Application of Methodologies Learned: Scenario Building, Prototyping, Strategic Foresight, and Backcasting

With your focal issue/domain selected, you will now apply a synthesized toolkit of futures methods to holistically model and envision alternative trajectories that space could take. This immersive process will weave together approaches like

Scenario Building

You will construct a set of multimedia scenario artifacts and experiential narratives depicting how different future realities could emergently unfold within your chosen issue space over longer timelines. These visceral scenarios will vividly illuminate key forces like technological impacts, sociocultural shifts, policy dynamics, and other factors shaping the possibilities.

For example, if exploring distributed education futures, you may build out scenarios spanning paradigms like

- *Decentralized Web3 skill ecosystems supplanting institutions*
- *Ubiquitous AI tutoring and credentialing models*
- *Biotech augmenting cognitive development and neural knowledge transfer*
- *Radical unschooling movements restructuring how we learn*

Prototyping Techniques

But you will not just document scenarios, you will contextualize them through multimedia prototypes and resonant artifacts that open immersive windows into inhabiting these futures. Leveraging generative

CHAPTER 9   SYNTHESIZING FUTURES PRACTICE

AI, VR/AR, experience design and more, you will craft visceral experiential embodiments that unlock felt-sense empathy for each scenario's ontological premises.

For the education futures, this could look like

- *Simulated "curriculum DAOs" modeling governance and incentive structures*
- *Resonant prototypes interfacing with affective AI tutoring systems*
- *Immersive VR environments depicting neurotechnology-enabled learning flows*
- *Speculative artifacts reflecting new skill signaling/ credentialing models*

Strategic Foresight

With scenarios and prototypes established, you will then employ strategic foresight frameworks to map potential impacts, emerging threats/opportunities and transition pathways related to each scenario through an organizational/institutional lens.

Methods like environmental scans, trend modeling, systems mapping, and participatory forecasting will reveal strategic considerations, partnership strategies, policy needs, and key uncertainties surrounding each prototyped education future.

You will derive strategic playbooks and decision-support systems incorporating probabilistic data, expert insights, and systems modeling to futureproof strategies against risks while creatively choreographing vision delivery.

Backcasting

Finally, you will apply the ethos of backcasting and construct comprehensive transformation roadmaps mapping out the policies, investments, innovations, and interventions required to progressively transition reality toward actualizing your selected preferred future scenario over the decades ahead.

CHAPTER 9   SYNTHESIZING FUTURES PRACTICE

This involves conducting root cause analysis to identify systemic acupuncture points requiring intervention, and orchestrating an integral "solutions ecology" of technology roadmaps, business models, cultural artifacts, policy packages and more that phase reality into embodying that regenerative vision through coordinated systemic redesign.

By integrating these methods into a cohesive applied practice, you will create a comprehensive futures territory for your focal issue, bridging the inspiration and pragmatism, the speculative and evidence-based, into a multifaceted body of insights, strategies, and transition pathways.

The goal is to demonstrate mastery in contextualizing the methods while fruitfully navigating the tensions between imagination and analytical rigor. To create artifacts that spark insight and awaken new potentials, while deriving pragmatic, accountable transformation roadmaps to start ushering in those visions as new realities.

Project Example: RideNow

> **Project background:**
>
> This chapter uses RideNow, a fictional composite of a large North-American ride-hailing operator, as a didactic example to show how strategic foresight can guide product and policy choices. The focal period is 2026–2032, with a 2023–2025 evidence baseline drawn from public sources (e.g., IEA mobility/EV reports, BNEF, city transport stats, regulatory filings).
>
> **How to read this example:**
>
> This playbook walks step-by-step through developing a foresight strategy for RideNow using purpose statements, STEEP+V/PESTLE scanning (rather than SWOT for 360 analysis), sensemaking, scenario building, prototyping, and backcasting.

CHAPTER 9   SYNTHESIZING FUTURES PRACTICE

Each stage illustrates how a mobility platform could explore plausible trajectories, monitor indicators, and translate insights into options, triggers, and near-term experiments.

In this project, we explore RideNow, a leading ride-sharing service, as a didactic case study to illustrate how strategic foresight methodologies can be applied to anticipate and shape the future of an industry. As the transportation landscape evolves rapidly, companies like RideNow must stay ahead of emerging trends, technological advancements, and shifting consumer expectations to maintain their competitive edge.

When working on taking a look at the future, it is essential to define the specific angles that will be under study and who will be working with these results, in what ways, which directly relates with what is the purpose of this particular study of the future.

This is why questioning will be registered in a purpose statement so that any stakeholder of an interested party can explore the work without losing context. For example, what is the future of plant-based processed foods? So in this case, we would not be looking at the general foods industry, but plant based, and not all plant based but processed ones, giving more focus to the study.

The purpose statement also considers the time factor. Is it mid term? Long term? Foresight works intent to look in a span of 10-15-20 years ahead. Even though that time frame gives room for potential preparations for those scenarios that could be disruptive or clearly changing the direction of how things are done in the present, we might also participate in the change as it is happening, helping it produce positive outcomes.

This reflection is important before we begin projects of this nature because although we cannot change the overall outlook of the future, we can from our individual efforts act in the most conscious, considerate, and responsible ways so that the future also becomes the result of our communal efforts toward a kinder and more elevated humanity in all the aspects of life.

CHAPTER 9   SYNTHESIZING FUTURES PRACTICE

9.2.5.1 Purpose Statement

Case scope: RideNow is a fictional composite of a large North-American ride-hailing operator. The example focuses on 2026–2032, using a 2023–2025 evidence baseline. To keep the context tangible, assume US Sunbelt metros (e.g., Phoenix, Dallas, Miami) with rapid growth, extreme-heat risk, and active EV policy.

Stakeholders: Riders (price-sensitive, reliability-seeking), drivers (earnings and safety), city agencies, utilities/charging providers, and neighborhood groups.

Constraints and realities: Tight unit economics, evolving labor regulation, heat-related service disruptions, charging availability, insurance risk, equity targets.

Purpose (Final statement):

"Design a medium-term foresight strategy (2026–2032) that enables RideNow to deliver reliable, lower-emission, heat-resilient mobility in Sunbelt metros while improving driver earnings and unit economics. The work will surface plausible trajectories, define no-regret moves and option bets, and specify indicators and trigger thresholds for switching among strategies."

How to craft a purpose statement (quick checklist):

1. Bound the context: Name the actor(s), geography, and time horizon.

2. Name the change focus: What must improve/transform (e.g., reliability, emissions, economics).

3. Stakeholders and Constraints: Call out the key groups and real-world constraints.

4. Success Measures: 2–5 initial KPIs and any equity/ethics guardrails.

CHAPTER 9   SYNTHESIZING FUTURES PRACTICE

Draft ➤ review with stakeholders ➤ compress to one clear sentence; log assumptions and version history.

The initial manual process would look like this:

Version 1: How can RideNow anticipate changes in people's transportation needs, preferences for ride-sharing services, and expectations for urban mobility solutions?

Version 2: How will people's transportation habits and preferences for ride-sharing evolve, and what will they expect from urban mobility services in the future?

Version 3: How will people move around cities and use ride-sharing services in new ways, and what will their expectations be for urban transportation in 15 years?

Version 4: How will people use ride-sharing and move around cities in 2040, and what will they expect from urban transportation services?

So this is what is happening in these iterations:

Version 1 starts broad, encompassing the main areas RideNow needs to consider: Changes in transportation needs, ride-sharing preferences, and expectations for urban mobility.

Version 2 focuses more on the evolution of habits and preferences, making the statement more forward-looking.

Version 3 introduces a specific time frame (15 years) and emphasizes new ways of using services, aligning with the idea of anticipating future changes.

Version 4 further refines the statement by specifying the exact year (2040) and simplifies the language for clarity.

The iterative process narrows the focus of the purpose statement making it more specific and actionable as well as defining a specific timeline while simplifying the language for better understanding.

Now that we understand where this process comes from, we can use a generative AI tool; in this case I am using Claude from Anthropic to evaluate your initial attempts and provide suggestions for cocreating,

CHAPTER 9   SYNTHESIZING FUTURES PRACTICE

collaborating in the process. It is worth emphasizing that this process ideally should not be undertaken only by an AI tool but as a collaborative effort, refining and iterating.

How can I ask the AI platform to provide this support for a case like this you might ask? It is important when communicating with an AI assistant to provide as much background as possible about your project or concept to work with. In this case, you might create a summary where you write what is the general idea of the business or concept under study and the work intended. Then you begin adding more details so that the AI assistant systematically progresses through the rational process making the process more precise and efficient than producing many back and forth prompt iterations.

You might start the process with a prompt like this: "The following file contains information about –Insert business case or project–. Please read it and prepare a purpose statement for a strategic foresight study under the following framework: –Insert the framework you need it to work with–."

Then once you obtain the initial results you might discuss with the team members or stakeholders additional refinements to make it adjusted for the particular needs and nuances required for the project.

9.2.5.2 Foresight Framing

The purpose of this strategic foresight analysis is to explore and anticipate the future landscape of the ride-sharing industry, with the goal of positioning RideNow as a proactive leader in the evolving market. By examining emerging trends, potential disruptions, and shifting consumer behaviors, we aim to

1. Identify key drivers shaping the future of urban mobility and ride-sharing services.
2. Anticipate potential challenges and opportunities that may arise in the next 5–10 years.

CHAPTER 9   SYNTHESIZING FUTURES PRACTICE

3. Develop a set of plausible future scenarios for the ride-sharing industry.

4. Generate innovative ideas for new services, technologies, or business models that could give RideNow a competitive edge.

5. Inform the development of robust, forward-thinking strategic initiatives that will enable RideNow to not only adapt to future changes but to actively shape the industry's direction.

With that in mind, we proceed to develop our framework for this strategy:

Vision: What drives our foresight exploration? We need a compelling vision to guide our decisions and inspire our journey through the foresight process.

Stakeholders: Who will benefit from our insights? Understanding their backgrounds, expertise in futures thinking, and familiarity with our vision is key. We must also consider the depth of information they require.

Medium: How will we bring our foresights to life? Will it be an objective, detailed report, a thought-provoking presentation, or a catalyst for change?

Engagement: How will our stakeholders interact with our foresight output? Will they collaboratively explore its elements, or individually reflect on its implications?

Impact: What actions will our foresights inspire? Are we planting seeds for future consideration, or providing a blueprint for immediate action?

Let's put this into work by sharing this question in a generative AI platform (in this case we continue using Claude from Anthropic) so that we can evaluate the results and the framework.

As we have learned so far, it is important to offer context from the prompts that we write, so in this case we will be using the same initial dialogue that we start to offer continuity to the AI assistant so that it can

CHAPTER 9   SYNTHESIZING FUTURES PRACTICE

better produce results. It is recommended that in this step we engage some team members and stakeholders first to offer context to the AI assistant and then proceed with the use of its reasoning. It is because there are often varied nuances in a business that only the human context can have through past experiences, and those important discussions might have important value to the determination of the framework to ask the right questions.. Once those inputs are defined, we can proceed to generate versions until we feel closer with a more reflective framework for the nature of the study.

After those deliberations, initial AI cocreation, and human review, this is how our final version looks like:

Vision

How will people move around cities and use ride-sharing services in 2040, and what will be their expectations of urban transportation?

Stakeholders

- RideNow executive leadership team
- Department heads
- Product owners
- Innovation and insights teams
- Strategic planning group

Medium

Strategic playbook for urban mobility futures, including

- Comprehensive written report
- Interactive digital presentation
- Scenario-based workshop materials

Engagement

Provide a roadmap to anticipate possible futures of urban transportation and potential strategic responses for RideNow.

Impact

- Identify and reframe emerging opportunities and challenges in urban mobility
- Guide strategic decision-making and innovation initiatives
- Inform long-term investment and partnership strategies

Once we have passed those iterations through human discussion and proactive participation and engagement and the stakeholders are confident on the proposed framework, we can move to the next stage of the study: The 360 Analysis.

9.2.5.3 360 Analysis (External STEEP+V Scan)

We begin with an external 360 scan using STEEP+V (Social, Technological, Economic, Environmental, Political+Values) to surface signals, drivers, and critical uncertainties that shape ride sharing. This scan prioritizes the few forces most likely to bend trajectories and informs all subsequent choices.

External drivers snapshot (examples):

> S: Safety perceptions, hybrid/remote work patterns, equity-of-access expectations

> T: AV pilots/permits, dispatch AI, EV charging density

> E: Driver earnings/unit economics, insurance costs

> E(nv): Extreme-heat days, clean-air zones, emissions targets

CHAPTER 9   SYNTHESIZING FUTURES PRACTICE

P/L: Labor classification, curb/congestion pricing, privacy rules

V: Sustainability norms, labor fairness, data trust

Output: 2–3 critical uncertainties, a tagged signal list, and early indicators.

We then translate these external findings into an internal diagnostic (SWOT) to align capabilities and constraints with the outside-in picture (see Figure 9-1).

As previously described, it is recommended to continue to use the same dialogue box with an AI assistant so that the context established about the company background and details as well as any market analysis is already understood and preserved. From that info, it can be complemented with any other particular detail required to know in terms of current circumstances around the present situation of the market and the company positioning. Then we might engage with the AI assistant to provide rational insights on the analysis, followed by the human review for accurate results and purpose.

Clarifying Scopes: In foresight, STEEP+V maps the external landscape. SWOT summarizes internal Strengths/Weaknesses and external Opportunities/Threats. We first scan outside-in (STEEP+V), then reflect inside-out (SWOT) to see how the organization can respond.

9.2.5.4 STEEP+V Analysis

As we delve deeper into our foresight study for RideNow, our next crucial step is to identify and analyze key signals that will shape the future of ride-sharing. To ensure a comprehensive view, we'll employ the STEEP+V model, which will help us create a 360° perspective of the forces influencing RideNow's business landscape.

CHAPTER 9   SYNTHESIZING FUTURES PRACTICE

Why STEEP+V Analysis for RideNow?
The ride-sharing industry is at the intersection of technology, urban development, and changing social norms. By using STEEP+V analysis, we can capture a wide range of signals that go beyond just technological advancements, providing a holistic view of the forces shaping RideNow's future.

From Abstract Trends to Concrete Narratives
After identifying key signals using the STEEP+V framework, we'll introduce techniques to transform these abstract trends into concrete narratives. This step is crucial for making our foresight work accessible and relatable to all stakeholders at RideNow.

These narratives will help

1. *Illustrate potential future scenarios for ride-sharing*
2. *Highlight implications for RideNow's business model*
3. *Inspire innovative thinking among RideNow's teams*

Our Approach

1. *Conduct a thorough STEEP+V analysis, focusing on signals relevant to ride-sharing*
2. *Prioritize the most impactful and probable signals*
3. *Develop concrete narratives around key signals*
4. *Link these narratives to potential opportunities and challenges for RideNow*

Using the STEEP+V method will provide an even more comprehensive analysis for RideNow. STEEP+V stands for Social, Technological, Economic, Environmental, Political, and Values-based factors. This method will allow us to capture a broader range of signals, including environmental and values-based aspects that are particularly relevant to the evolving landscape of ride-sharing and urban mobility.

CHAPTER 9   SYNTHESIZING FUTURES PRACTICE

Let's reframe our introduction and approach using the STEEP+V method (following the same methodology in our last example, "the future of roads and highways" keeping in mind a collaborative approach with stakeholders to make sure biases are considered.)

Mapping the Future of Ride-Sharing: A STEEP+V Analysis for RideNow

Introduction to STEEP+V Analysis

As we delve deeper into our foresight study for RideNow, we are employing the STEEP+V method to identify and analyze key signals that will shape the future of ride-sharing. STEEP+V provides a comprehensive framework to create a 360° view of the forces influencing RideNow's business landscape.

This method is particularly suited for the ride-sharing industry, which sits at the intersection of technology, urban development, environmental concerns, and changing social values.

Why STEEP+V for RideNow?

1. *Comprehensive Coverage: STEEP+V ensures we don't overlook crucial factors like environmental impact and shifting societal values, which are increasingly important in the mobility sector.*

2. *Interconnected Analysis: It allows us to explore how different factors interact, providing a more nuanced understanding of potential futures.*

3. *Alignment with Sustainability Goals: The inclusion of environmental factors aligns with growing concerns about sustainable urban transportation.*

4. *Values-Based Insights: Understanding changing values helps predict shifts in consumer behavior and regulatory landscapes.*

CHAPTER 9   SYNTHESIZING FUTURES PRACTICE

Our STEEP+V Approach for RideNow

1. *Social Signals*
   - *Changing urban demographics*
   - *Evolving attitudes toward car ownership*
   - *Shifts in work patterns (e.g., remote work trends)*

2. *Technological Signals*
   - *Advancements in autonomous vehicle technology*
   - *Innovations in electric and alternative fuel vehicles*
   - *Development of smart city infrastructure*

3. *Economic Signals*
   - *Trends in the gig economy*
   - *Changes in urban real estate (parking spaces, land use)*
   - *Shifts in consumer spending on transportation*

4. *Environmental Signals*
   - *Climate change impacts on urban planning*
   - *Regulations on emissions and sustainable transportation*
   - *Trends in renewable energy adoption*

5. *Political Signals*
   - *Regulatory changes affecting ride-sharing and gig workers*
   - *Urban transportation policies*
   - *Data privacy and security legislation*

CHAPTER 9   SYNTHESIZING FUTURES PRACTICE

6. *Values-Based Signals*
   – *Growing emphasis on sustainability and eco-friendly choices*
   – *Changing perceptions of privacy and data sharing*
   – *Shifts in work–life balance priorities*

From Signals to Narratives
After identifying key signals using the STEEP+V framework, we will transform these insights into concrete narratives. This process will involve

1. *Prioritizing the most impactful and probable signals*
2. *Developing scenario narratives that illustrate potential futures*
3. *Linking these narratives to specific opportunities and challenges for RideNow*

These narratives will help

– *Illustrate complex interactions between different factors*
– *Make abstract trends tangible and relatable for all RideNow stakeholders*
– *Inspire innovative thinking and strategic planning*

Expected Outcomes
By the end of this STEEP+V analysis, we aim to provide RideNow with

1. *A comprehensive map of signals shaping the future of ride-sharing*
2. *A set of compelling future scenarios for the urban mobility landscape*
3. *Actionable insights to guide strategic decision-making*

CHAPTER 9   SYNTHESIZING FUTURES PRACTICE

4. *A foundation for developing innovative services and business models*

This analysis will equip RideNow with the foresight needed to navigate the complex and rapidly evolving world of urban mobility, positioning the company to thrive in various potential futures.

To proceed with this analysis, we would need to

1. *Conduct extensive research into each STEEP+V category, focusing on trends and signals relevant to ride-sharing and urban mobility.*

2. *Identify specific, concrete signals within each category.*

3. *Analyze how these signals might interact and influence each other.*

4. *Develop compelling narratives or scenarios based on the most significant signals.*

5. *Link these narratives directly to potential strategies and innovations for RideNow.*

Some methods to bring insight to the STEEP+V analysis

- *Start with you and how you interact with that service, product, or solution*

- *Use a tool like "A day in the life" to analyze when individuals interact with and use the solution and in what ways*

- *Ask people you know about their own uses of it*

- *Study similar or adjacent businesses and use them as analogies to enrich your own analysis and extract what have they done and what might they be working right now towards future plans*

CHAPTER 9   SYNTHESIZING FUTURES PRACTICE

- *Keyword searches and what counteracting forces exist for this business or solution*

Internal Diagnostic (SWOT)—RideNow (Illustrative)

Based on the STEEP+V scan above, summarized internal capabilities and constraints against external opportunities/threats are presented in Figure 9-1.

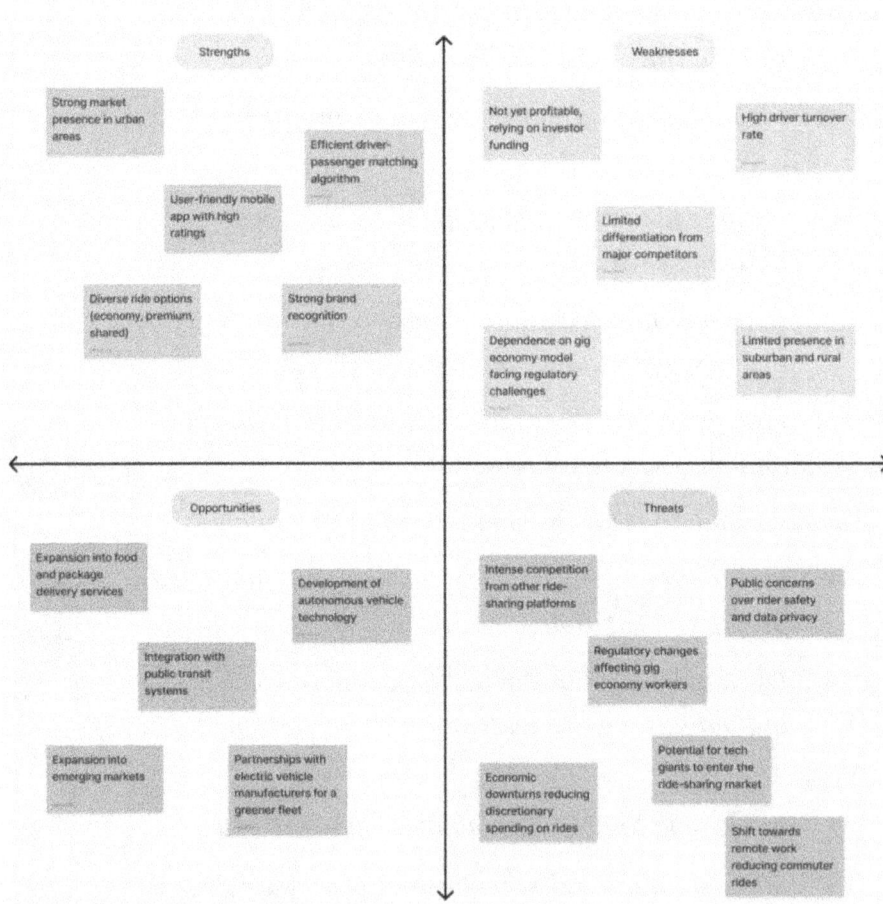

Figure 9-1. SWOT Internal Diagnostic RideNow (Illustrative)

158

CHAPTER 9   SYNTHESIZING FUTURES PRACTICE

Making informed research to find out this information is crucial. This is done through online consultations, interviews with stakeholders, team discussions, public engagement if that is the case and any other source of internal and external information that can be relevant for the study.

9.2.5.5 Collecting Signals

This step is crucial for our futures analysis. It involves collecting any type of information we see in our surroundings that we realize might have an impact directly or indirectly in our topic at hand. In this case, all kinds of information pertaining to ride hailing or ride sharing, other unexpected trends, transportation news, companies that affect this industry recent decisions, thoughts of the community we hear about this service, statistics and data recently shared, in summary any kind of data that might have an influence on our purpose statement. In our case, we have collected signals in a Figma board (See Figure 9-2 below), over a period of time to proceed then to analyze and categorize the data and extract insights from it based on the categorization in STEEP+V.

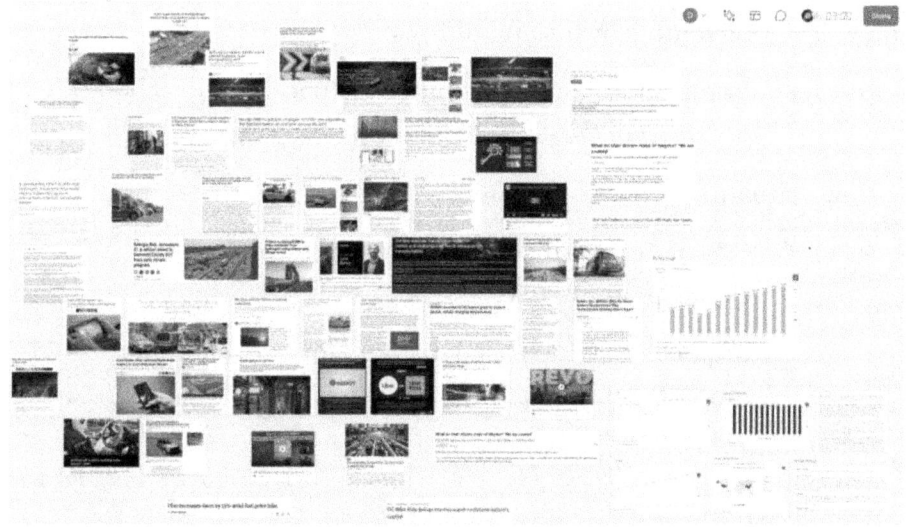

Figure 9-2. *Signals Collection Using Figma*

CHAPTER 9   SYNTHESIZING FUTURES PRACTICE

After we have performed the categorization of our data, these are some initial insights we have been able to infer and propose:

Social Signals

1. Aging Population in Urban Areas:
   - By 2035, 23% of the US population is projected to be over 65.
   - Implications: Increased demand for accessible, door-to-door transportation services.

2. Gen Z's Preference for Access Over Ownership:
   - 70% of Gen Z prefer using ride-sharing services over owning a car.
   - Implications: Potential for long-term growth in ride-sharing user base.

3. Remote Work Normalization:
   - 25% of all professional jobs in North America will be remote by the end of 2024.
   - Implications: Shift in commuting patterns, potential decrease in regular ride-sharing use.

4. Urbanization Trends:
   - 68% of the world population projected to live in urban areas by 2050.
   - Implications: Increased demand for efficient urban mobility solutions.

CHAPTER 9   SYNTHESIZING FUTURES PRACTICE

Technological Signals

1. Autonomous Vehicle Advancements:
   - Waymo's autonomous vehicles have driven over 20 million miles on public roads.
   - Implications: Potential for driverless ride-sharing fleets, reducing operational costs.

2. 5G Network Rollout:
   - 5G is expected to cover 65% of the world's population by 2025.
   - Implications: Enhanced real-time tracking, improved in-car entertainment options.

3. Artificial Intelligence in Ride Matching:
   - AI algorithms reduce wait times by up to 30% in pilot programs.
   - Implications: Improved efficiency and user experience.

4. Electric Vehicle (EV) Adoption:
   - EVs are projected to account for 31% of all cars on the road by 2040.
   - Implications: Pressure to transition to electric fleets, potential for reduced operational costs.

Economic Signals

1. Gig Economy Growth:
   - Gig economy is projected to reach $455 billion globally by 2023.
   - Implications: Continued access to flexible workforce, but potential regulatory challenges.

2. Micromobility Market Expansion:
   - Global micromobility market is expected to reach $300 billion by 2030.
   - Implications: Opportunities for integration with e-scooters and bike-sharing services.

3. Urban Parking Space Repurposing:
   - Cities like San Francisco are converting 10% of parking spaces to other uses by 2025.
   - Implications: Potential for dedicated ride-sharing pickup/drop-off zones.

4. Subscription-Based Transportation Models:
   - Subscription services in the mobility sector growing at 40% annually.
   - Implications: Opportunity for RideNow to develop subscription-based offerings.

Environmental Signals

1. Carbon Neutrality Targets:
   - Over 100 countries pledged to carbon neutrality by 2050.
   - Implications: Increased pressure to adopt zero-emission vehicles and sustainable practices.

2. Urban Air Quality Regulations:
   - Cities like London implementing Ultra Low Emission Zones.
   - Implications: Potential restrictions on high-emission vehicles, favoring electric ride-sharing fleets.

CHAPTER 9 SYNTHESIZING FUTURES PRACTICE

3. Extreme Weather Events:

    - Climate-related disasters increased by 83% in the last 20 years.
    - Implications: Need for resilient transportation options during emergencies.

4. Circular Economy in Transportation:

    - Global market for remanufactured auto parts is expected to reach $100 billion by 2025.
    - Implications: Opportunities for sustainable vehicle maintenance and recycling programs.

Political Signals

1. Gig Worker Classification Laws:

    - California's AB5 law reclassifying gig workers as employees.
    - Implications: Potential increase in operational costs, need for new business models.

2. Data Privacy Regulations:

    - GDPR-like regulations being adopted globally.
    - Implications: Need for robust data protection measures, potential limitations on data use.

3. Smart City Initiatives:

    - Over 1,000 smart city pilot projects worldwide.
    - Implications: Opportunities for integration with city transportation systems.

CHAPTER 9 SYNTHESIZING FUTURES PRACTICE

4. Autonomous Vehicle Regulations:

    - The US Department of Transportation releasing new AV guidelines.

    - Implications: Clearer path for autonomous ride-sharing implementation.

Values-Based Signals

1. Prioritization of Work–Life Balance:

    - 72% of workers prioritizing work–life balance in job decisions.

    - Implications: Changing patterns of transportation use, potential for leisure-focused ride services.

2. Growing Environmental Consciousness:

    - 73% of global consumers are willing to change habits to reduce environmental impact.

    - Implications: Increased demand for eco-friendly transportation options.

3. Trust and Safety Concerns:

    - 58% of US adults concerned about safety in ride-sharing.

    - Implications: Need for enhanced safety features and transparent practices.

4. Community-Centric Consumption:

    - 63% of global consumers prefer to buy from companies that support social causes.

    - Implications: Opportunity for community-focused initiatives and local partnerships

9.2.5.6 Synthesis

Our next step is to take the foundational elements from our analysis performed so far and start combining them in a fresh, creative way, allowing new possibilities to emerge from the data. By the conclusion of the synthesis process, we begin to see concrete ideas taking shape, ready to be tested and refined as you address any knowledge gaps that were uncovered (See Figure 9-3). Here we need to keep in mind that if we infer an idea from a category, we have to make sure that it has evidence and support from the data collected so that we perform the analysis and synthesis as objectively as possible. This is the step where biases can filter in and we need to make sure that the participatory process is solid and well implemented so that not only the idea of a consultant or designer is in place, but that potential insights are coming from a larger web of thoughts and participants as strong insights.

To make sure we are still on track and aligned with the purpose statement, we need to reverse engineer to monitor and check this correlation. This can be done by assessing if the groups of insights found clearly align with the purpose statement by creating a mapping check from idea to data to situational analysis to the purpose as illustrated below.

After we have done our check, it is recommended to validate those findings with stakeholders and interested parties to make sure that our data is producing real and valuable direction for potential course of actions. After this feedback loop, we revise our board (the Figma board with the collected signals from this example) and further refine, confirm, and validate the classification of the signals that we have collected to start unveiling objective trends in our data.

CHAPTER 9   SYNTHESIZING FUTURES PRACTICE

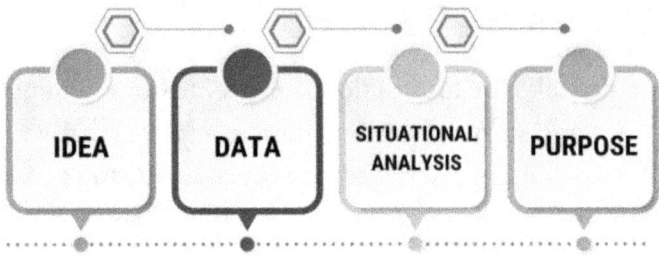

Figure 9-3. Mapping Back to Purpose

9.2.5.7 Speculation/Hypothesis

In this step, we transform those found ideas or trends into speculations to create some perspective. Here is where we begin to see and state a relationship of those insights with our statement question, with our business or case study.

To create the hypothesis, we will group out the ideas we have from the synthesis step. We can create between 5 to 10 of them to eventually begin unveiling some actions that RideNow needs to take.

For each speculation, we will look into giving it a name, something like a branding identifier and a description of what might happen in that speculation, developing like this our first layer of analysis.

Once this process is done, we move onto creating the anticipations where we talk about the how, for each of the speculations or hypothesis, analyzing the impacts on the customer, impacts on the business (what do we need to change, the impact in the current vs. potential future structure, capabilities as a company, impact on the products and services we deliver, who do we partner with, etc.), and what we need to do to respond to it (what specific steps might be required to begin preparing). In a practical sense, this can be created in a matrix that can be created and analyzed manually with traditional pen and paper or as sophisticated as using an AI assist to help with several iterations making sure all the findings are aligned with the purpose statement.

CHAPTER 9   SYNTHESIZING FUTURES PRACTICE

The speculations are intended to give us a hint to what might happen in those different scenarios and offer a potential direction on how the business, RideNow, needs to adapt and strategize accordingly to arrive at what eventually will become a roadmap for the company. In our example, the speculations and anticipations matrix might look like this:

Speculation Name: Embrace autonomous ride-sharing

Description: Transition to a fleet of self-driving vehicles for ride-sharing services

Impact on Customer:

- Potentially lower fares due to reduced operational costs
- Increased safety with elimination of human error
- Possible reduction in personal interaction

Impact on Business:

- Significant reduction in operational costs
- Potential loss of current driver workforce
- Large initial investment in technology and vehicles

Potential Actions:

- Partner with leading autonomous vehicle manufacturers
- Develop retraining programs for current drivers
- Create new roles for human oversight and customer service

Speculation Name: Provide multimodal mobility solutions

Description: Integrate ride-sharing with other forms of transportation (e.g., e-scooters, public transit)

CHAPTER 9   SYNTHESIZING FUTURES PRACTICE

Impact on Customer:

- Seamless door-to-door transportation options
- Potential cost savings through efficient combinations

Impact on Business:

- Expanded service offerings and revenue streams
- Increased complexity in operations and partnerships

Potential Actions:

- Develop partnerships with public transit authorities and micromobility companies
- Create a unified app for multimodal transportation planning

Speculation Name: Offer sustainable transportation options
Description: Transition to an all-electric or zero-emission vehicle fleet
Impact on Customer:

- Eco-friendly transportation choices
- Potential for improved air quality in urban areas

Impact on Business:

- Alignment with environmental regulations and consumer values
- Potential for government incentives
- Higher initial costs for vehicle acquisition

Potential Actions:

- Gradually replace fleet with electric vehicles
- Invest in charging infrastructure
- Develop partnerships with EV manufacturers

CHAPTER 9 SYNTHESIZING FUTURES PRACTICE

Speculation Name: Implement dynamic urban space utilization

Description: Use data to optimize pickup/drop-off locations and create dedicated spaces.

Impact on Customer:

- Reduced wait times and more convenient pickup points
- Potentially safer boarding and alighting

Impact on Business:

- Improved efficiency in operations
- Better relationships with city planners and officials

Potential Actions:

- Collaborate with city governments on smart city initiatives
- Develop AI-driven systems for real-time space optimization

Speculation Name: Personalize the ride experience

Description: Use AI and data analytics to tailor rides to individual preferences

Impact on Customer:

- More enjoyable and personalized ride experiences
- Potential privacy concerns

Impact on Business:

- Increased customer loyalty and satisfaction
- Differentiation from competitors

CHAPTER 9   SYNTHESIZING FUTURES PRACTICE

Potential Actions:

- Invest in AI and machine learning capabilities
- Develop strict data privacy and security measures

Speculation Name: Create community-centric ride-sharing
Description: Develop features that foster local connections and support local businesses
Impact on Customer:

- Sense of contributing to local community
- Discovery of local businesses and events

Impact on Business:

- Improved brand image and customer loyalty
- New partnership opportunities with local businesses

Potential Actions:

- Implement features for discovering and supporting local businesses
- Create community-focused marketing campaigns

Speculation Name: Provide flexible work arrangements
Description: Develop new models for driver engagement and compensation
Impact on Customer:

- Potential for more diverse driver pool and availability
- Possible fluctuations in service quality

Impact on Business:

- Improved driver retention and satisfaction
- Potential regulatory challenges

Potential Actions:

- Experiment with various work models (e.g., shifts, part-time options)
- Engage with policymakers on gig economy regulations

Speculation Name: Enhance ride safety and trust

Description: Implement advanced safety features and transparent practices

Impact on Customer:

- Increased peace of mind and trust in the service
- Potential for slight increase in fares

Impact on Business:

- Improved brand reputation and customer loyalty
- Potential reduction in liability and insurance costs

Potential Actions:

- Invest in advanced safety technologies (e.g., in-car cameras, real-time monitoring)
- Develop comprehensive safety training programs

9.2.5.8 Strategy Building

With the Anticipation and Speculation matrix now complete, the next step is to prioritize and categorize the speculations in order to outline three distinct strategic paths. For each strategy, we will create a scenario—these will be narrative-driven, focusing on people and brought to life through text, visuals, or graphics to paint clear and compelling pictures. These scenarios will highlight gaps in each strategy that need to be addressed. Lastly, we will establish evaluation criteria to assess the three strategies and determine the best one for the business.

CHAPTER 9   SYNTHESIZING FUTURES PRACTICE

Potential Strategy 1: Autonomous Mobility Network

Name: Autonomous Mobility Network

Description: By 2035, RideNow has transitioned to a fully autonomous fleet, integrated seamlessly with other forms of public and private transportation. Our AI-driven system optimizes routes, reduces traffic congestion, and provides personalized travel experiences. Users simply input their destination, and our network determines the most efficient combination of autonomous vehicles, public transit, and micro-mobility options to get them there.

What the strategy means for your customer: Customers enjoy hassle-free, door-to-door transportation without the need to own a vehicle or navigate complex transit systems. They benefit from reduced travel times, lower costs, and increased safety. The personalized in-vehicle experience caters to individual preferences, whether for work, relaxation, or entertainment.

What strategy means for the business: RideNow transforms from a ride-sharing company to a comprehensive mobility provider. We see significant reductions in operational costs due to the elimination of human drivers, offset by increased technology investments. Our data-driven approach allows for predictive maintenance and optimal fleet management.

Differentiation: RideNow's early adoption of autonomous technology and our unique integration with other transportation modes position us as the leader in smart urban mobility. Our personalized user experiences set us apart from competitors who may offer similar services.

Mid-term view: By 2030, 50% of our fleet is autonomous in major urban areas, with human-driven vehicles still available in less densely populated regions. We've established key partnerships with public transit authorities and micro-mobility providers in several cities.

Long-term view: By 2035, our fully autonomous network spans across urban and suburban areas. We've expanded into logistics, offering autonomous delivery services. Our AI system has become an integral part

CHAPTER 9   SYNTHESIZING FUTURES PRACTICE

of smart city infrastructure, helping to optimize overall traffic flow and urban planning.

Potential barriers: Regulatory hurdles in some regions may slow adoption. Public trust in autonomous vehicles needs to be built. High initial investment costs and potential cybersecurity threats need to be managed.

Immediate steps to take:

1. Forge partnerships with leading autonomous vehicle manufacturers.
2. Initiate pilot programs in receptive cities.
3. Invest heavily in AI and machine learning capabilities.
4. Engage with policymakers to shape favorable regulations.
5. Develop a comprehensive public relations strategy to build trust in autonomous technology.

Potential Strategy 2: Community-Centric Mobility Platform
Name: Community-Centric Mobility Platform
Description: By 2035, RideNow has evolved into a hyperlocal, community-focused mobility platform. We facilitate not just rides, but connections within communities. Our platform supports local businesses, enables community events, and fosters neighborhood relationships, all while providing sustainable transportation options tailored to each unique community's needs.

What the strategy means for your customer: Customers see RideNow as more than a ride-sharing service; it's a gateway to their community. They can discover local events, support neighborhood businesses, and connect with fellow community members. Transportation becomes a social experience that strengthens local ties.

173

CHAPTER 9   SYNTHESIZING FUTURES PRACTICE

What strategy means for the business: RideNow diversifies its revenue streams through partnerships with local businesses and community organizations. We become deeply embedded in the fabric of each community we serve, increasing customer loyalty and expanding our user base beyond traditional ride-sharing demographics.

Differentiation: While competitors focus on efficiency and speed, RideNow emphasizes community building and local economic development. Our deep community integration creates a unique value proposition that's difficult for others to replicate.

Mid-term view: By 2030, we've launched community hubs in major cities, partnering with local businesses to offer exclusive deals and facilitate community events. Our ride-sharing service includes options for shared rides with community members heading to the same local events.

Long-term view: By 2035, RideNow is recognized as a crucial community infrastructure in cities worldwide. We've expanded to include community-focused features like neighborhood watch programs, local crowdfunding initiatives, and community-driven urban development projects.

Potential barriers: Building strong community ties takes time and varies greatly between locations. Balancing global scale with local focus could be challenging. Some communities may resist the level of integration we're proposing.

Immediate steps to take:

1. Identify pilot communities for initial rollout.
2. Develop partnerships with local businesses and community organizations.
3. Create a team focused on community engagement and local adaptation of our platform.
4. Invest in AI capabilities to personalize community recommendations.

CHAPTER 9   SYNTHESIZING FUTURES PRACTICE

5. Design and implement community feedback mechanisms to continually improve our local relevance.

Potential Strategy 3: Sustainable Mobility Ecosystem
Name: Sustainable Mobility Ecosystem

Description: By 2035, RideNow leads the transition to fully sustainable urban transportation. Our all-electric, solar-powered fleet is complemented by an extensive network of green micro-mobility options. We've integrated circular economy principles into our operations, from vehicle manufacturing to end-of-life recycling. Our platform incentivizes sustainable choices and actively contributes to reducing cities' carbon footprints.

What the strategy means for your customer: Customers can travel with zero guilt about their environmental impact. They earn sustainability points for choosing eco-friendly options, which can be redeemed for rewards or donated to environmental causes. Our app provides real-time information about the positive environmental impact of their transportation choices.

What strategy means for the business: RideNow positions itself as the go-to platform for environmentally conscious consumers and cities aiming to meet ambitious climate goals. We benefit from government incentives for green initiatives and attract environmentally focused investors. Our circular economy approach reduces long-term costs and resource dependencies.

Differentiation: While others may offer some green options, RideNow's comprehensive approach to sustainability across our entire ecosystem sets us apart. We're not just a transportation company, but a key player in the global fight against climate change.

Mid-term view: By 2030, our fleet is 75% electric in major cities. We've established solar charging stations and battery swapping infrastructure. Our app includes features that allow users to track and offset their carbon footprint across all transportation modes.

175

Long-term view: By 2035, we've achieved carbon-negative operations through our sustainable practices and reforestation initiatives. Our model has been adopted by cities worldwide as a blueprint for sustainable urban mobility. We've expanded into sustainable urban planning consulting, leveraging our expertise and data.

Potential barriers: High initial costs for transitioning to a fully sustainable fleet. Variations in government support for green initiatives across different regions. Potential shortages of raw materials for electric vehicle batteries.

Immediate steps to take:

1. Accelerate the transition to an electric fleet, starting in the most environmentally conscious cities.

2. Develop partnerships with renewable energy providers and electric vehicle manufacturers.

3. Invest in research and development for sustainable vehicle technologies.

4. Create a sustainability task force to oversee the implementation of circular economy principles.

5. Launch marketing campaigns to educate users about the environmental impact of their transportation choices.

9.2.5.9 Scenario Building

In this step, we might brainstorm the best suitable scenario tool or medium to represent these strategies. To help with their stories, you can use a platform like Claude to help brainstorm ideas for suitable scenarios and engage stakeholders for final decision making. Once this process is

CHAPTER 9   SYNTHESIZING FUTURES PRACTICE

complete, for our case study we have used "a day in the life" to obtain a closer look at how the dynamics of the customer might unfold under our particular speculations.

These scenarios provide a more emotionally engaging and relatable perspective on the three potential strategies we developed for RideNow.

Here's a brief analysis of each scenario:

1. **Autonomous Mobility Network**: This scenario emphasizes the seamless, efficient, and personalized experience of using a fully integrated autonomous transportation system. It highlights how this technology can reduce stress, increase productivity, and adapt to the user's needs throughout the day.

2. **Community-Centric Mobility Platform**: This scenario showcases how RideNow can become more than just a transportation service, fostering community connections, supporting local businesses, and facilitating civic engagement. It illustrates the emotional benefits of feeling more connected to one's neighborhood and community.

3. **Sustainable Mobility Ecosystem**: This scenario demonstrates how RideNow can make sustainability tangible and rewarding for users. It shows how eco-friendly transportation choices can be seamlessly integrated into daily life, providing users with a sense of accomplishment and contribution to a larger cause.

RideNow 2035: A Day in the Life Scenarios
Scenario 1: Autonomous Mobility Network
Meet Sarah, a 32-year-old marketing executive living in Metro City.

CHAPTER 9   SYNTHESIZING FUTURES PRACTICE

7:00 AM: Sarah wakes up and checks her RideNow app. It suggests leaving in 20 minutes for her 9 AM meeting across town, factoring in current traffic and her preference for a quick coffee stop.

7:50 AM: As Sarah steps out, a sleek, autonomous RideNow vehicle pulls up. The door opens automatically, greeting her by name. Inside, the temperature and lighting are set to her preferences.

8:05 AM: The vehicle smoothly merges into a dedicated autonomous lane. Sarah uses the in-car workstation to review her presentation while the vehicle's AI assistant provides a briefing on her day's schedule.

8:20 AM: The vehicle detours slightly to Sarah's favorite coffee shop. Her usual order is waiting as she steps out for a quick pickup.

8:55 AM: Sarah arrives at her meeting location, feeling relaxed and prepared. The RideNow app has already scheduled her return trip, optimizing for her afternoon appointments.

6:30 PM: After work, Sarah uses RideNow to meet friends for dinner. The app suggests a shared ride with a colleague heading in the same direction, saving costs and reducing emissions.

10:00 PM: Returning home, Sarah reflects on how seamless and stress-free her day has been, thanks to RideNow's integrated autonomous mobility network.

Scenario 2: Community-Centric Mobility Platform

Meet Carlos, a 45-year-old teacher and active community member in Harmony Heights.

7:30 AM: Carlos opens his RideNow app to plan his day. He notices a community alert about a neighborhood cleanup event this weekend and signs up to volunteer.

8:00 AM: Carlos boards a RideNow shared shuttle to school. He chats with his neighbors, learning about a new community garden project from a fellow passenger.

3:30 PM: After school, Carlos uses RideNow to take a group of students to the local science museum. The app automatically applies an education discount and suggests relevant exhibits.

5:00 PM: On the way home, Carlos stops at a local bakery he discovered through RideNow's community highlights feature. He enjoys chatting with the owner, feeling more connected to his neighborhood.

7:00 PM: Carlos attends a community meeting about local development. He used RideNow's community polling feature to vote on discussion topics beforehand.

9:00 PM: Returning home, Carlos feels a strong sense of belonging. He appreciates how RideNow has transformed from a simple ride-sharing app into a platform that strengthens community bonds and supports local businesses.

Scenario 3: Sustainable Mobility Ecosystem
Meet Zoe, a 28-year-old environmental scientist living in Green City.

6:00 AM: Zoe starts her day by checking her RideNow sustainability score. She's pleased to see she's in the top 10% of eco-friendly users in her city.

7:30 AM: Zoe uses the RideNow app to reserve an electric bike for her morning commute. The app shows her how much CO_2 she's saving compared to other transport options.

1:00 PM: For a field research trip, Zoe books a solar-powered RideNow vehicle. The app informs her that the vehicle's materials are 95% recycled and recyclable.

4:30 PM: Returning to the city, Zoe opts for a shared ride. The app matches her with passengers with similar environmental interests, leading to an engaging discussion about local sustainability initiatives.

6:00 PM: Zoe uses her accumulated sustainability points to "plant" a virtual tree in the RideNow forest. She sees that, collectively, RideNow users in her city have offset thousands of tons of CO_2 this month.

8:00 PM: At a dinner with friends, Zoe shares her RideNow impact report, sparking a conversation about personal contributions to fighting climate change.

10:00 PM: As she prepares for bed, Zoe feels empowered, knowing that her daily transportation choices with RideNow are making a real difference for the environment.

9.2.5.10 Strategy Selection

Now it's the time to use a matrix selection criteria to determine which option is the most suitable for a strategy in the RideNow case study.

To select the best strategic direction, we will use specific criteria. It's recommended to start with three broad criteria to ensure a balanced decision-making process:

1. **Confidence in speculations**—How likely is the scenario to occur?

2. **Customer impact**—How will it improve or add value to the customer's life?

3. **Business opportunity size**—Is the opportunity large enough to be profitable and attract a substantial customer base?

4. **Business feasibility**—How capable is your company of executing this strategy effectively?

Additional criteria can be added based on the business and strategy. For example, one could be **creating buzz and making a bold move**, especially if your brand already has a strong presence.

You can evaluate each strategy based on these criteria, either by listing pros and cons or through group voting, where participants justify their choices. You could also assign scores to each criterion and tally them up to find the most favorable strategy.

Remember, this process is a tool to guide your decision-making. If the outcome does not feel right, reconsider the criteria or review the strategies to ensure nothing is overlooked. In this example, we have used Claude from Anthropic to aid with the data analysis and visual generation as depicted below.

CHAPTER 9   SYNTHESIZING FUTURES PRACTICE

Let's take a look at the strategy decision matrix:

```
RideNow Weighted Decision Matrix - Text Format

Criteria                              Weight   AMN    CCP    SME

Confidence in Underlying Speculations   20%     9      7      8
Impact on Customer                      25%    10      8      7
Size of Business Opportunity            25%    10      7      8
Feasibility of RideNow to Respond       15%     8      9      7
Potential for Industry Leadership       15%    10      7      8

Total Weighted Score                   100%    9.5    7.55   7.6

AMN = Autonomous Mobility Network
CCP = Community-Centric Platform
SME = Sustainable Mobility Ecosystem

Scoring: 1-10 scale, 10 being highest
Weighted scores calculated but not shown for simplicity
```

Figure 9-4. *RideNow Strategy Decision Matrix*

RideNow: Strategy Selection and Elements to Keep
A. Robustness and Selection Rationale
After careful consideration of our three potential strategies (Autonomous Mobility Network, Community-Centric Mobility Platform, and Sustainable Mobility Ecosystem), we have selected the Autonomous Mobility Network as RideNow's primary future strategy.

How the Decision Was Made:
Rather than treating any one future as inevitable, we ran the three strategies through wind-tunneling against counterfactual conditions (regulatory drag, heat/infra stress, demand volatility, capital constraints). The Autonomous Mobility Network (AMN) performs well if safety, regulation, and infra reliability trend positively and unit economics clear. Where those assumptions fail, AMN under-performs—so we hold it as a conditional bet, not a foregone conclusion.

181

CHAPTER 9   SYNTHESIZING FUTURES PRACTICE

Competing Logistics We Considered (Counterfactuals):

- AV Stall: Safety incidents, labor rulings, and insurance costs slow deployment beyond 2030.

- Transit + Shared Renaissance: Policy favors transit/micromobility; priced curb access reduces solo AV trips.

- Climate and Infra Stress: Extreme heat reduces EV range/charger uptime; outages create service deserts.

Systemic Risks and Equity Flags (Cross-Scenario):
Worker displacement, curb/land-use externalities, privacy harms from in-cabin sensing, heat-day access gaps in low-income areas. These are addressed via guardrails and indicators below.

2026–2032 portfolio stance

- No-regret moves: Heat-day operations protocol; charger-uptime telemetry; neighborhood access SLOs; privacy-by-design defaults; driver transition fund, and re-skilling pilots.

- Real options (stage-gated): Limited AV pilots in two Sunbelt metros; depot charging with islandable power; AV insurance partnerships.

- Big bet (conditional): Scale AMN only if trigger thresholds are met for four consecutive quarters.

Indicators ➤ Trigger Thresholds (Examples):

- City AV incident rate ≤ 0.3 / 1M miles and trending down.

- Fast-charge uptime ≥ 92% metro average (including heat season).

- Net driver income ≥ local living-wage benchmark during transition.

- Public approval of AV services ≥ 60% in host-metro surveys.

Making Trade-offs Legible (Design Futures):
To surface frictions—not only benefits—the team prototypes and tests: a heat-advisory service blueprint; a curb-pricing "receipt"; a privacy-consent flow for in-cabin sensing (and a privacy-preserving variant); a driver transition notice; and a neighborhood access dashboard. Each artifact carries a test card (hypothesis ➤ evidence ➤ decision rule) and is used in red-team reviews and premortems to reduce utopian drift.

B. Elements Retained from Alternative Strategies
To avoid single-path bias, the evaluation retains high-value elements from the other strategies and assigns them to the portfolio as no-regret moves, real options, or guardrails.

From Community-Centric Mobility Platform

- Local partnerships (no-regret): Ongoing MOUs with community orgs and small businesses for curb access, hubs, and outreach.

- Community features (option): Shared rides to local events; neighborhood route optimization pilots.

- Hyperlocal customization (guardrail): Neighborhood access SLOs; publish wait-time dashboards by census tract.

From Sustainable Mobility Ecosystem

- Electric fleet transition (no-regret): Prioritize EVs in new leases; depot charging with islandable power.

- Sustainability metrics (guardrail): Per-trip CO_2e and charger-uptime KPIs; publish quarterly.

- Circular operations (option): Battery refurbishment/reuse pilots with suppliers.

- City collaboration (option): Data-sharing for clean-air zones and curb pricing trials.

By focusing on the Autonomous Mobility Network while keeping these elements in mind, RideNow can pursue a future-forward strategy that also incorporates community engagement and sustainability. This approach allows us to adapt our core strategy as market conditions and technologies evolve, ensuring we remain at the forefront of urban mobility solutions.

Strategic Foresight Playbook + Futures Design Playbook
As you might suspect by now, organizations need a structured approach to anticipate and prepare for future challenges and opportunities. This is where a Strategic Foresight Playbook becomes an invaluable tool.

A Strategic Foresight Playbook is a comprehensive document that outlines the organization's vision for the future and the strategic steps needed to achieve it. It is not just a static report but rather, it is a dynamic tool that combines insights from your foresight analysis with actionable plans, serving as a guide for decision-making and strategic planning.

Why is a Strategic Foresight Playbook necessary? There are several compelling reasons:

1. Clarity of Vision: It provides a clear, articulated vision of your preferred future, helping align all stakeholders toward a common goal.

2. Strategic Alignment: The playbook ensures that short-term actions and mid-term plans are in line with long-term objectives.

CHAPTER 9   SYNTHESIZING FUTURES PRACTICE

3. Adaptability: By considering multiple future scenarios, it helps your organization remain flexible and responsive to change.

4. Decision Support: It offers a framework for making strategic decisions, ensuring they are consistent with your long-term vision.

5. Communication Tool: The playbook serves as a powerful means to communicate your strategy across the organization and to external stakeholders.

6. Action Orientation: It transforms abstract foresight insights into concrete, actionable plans.

The Strategic Foresight Playbook is the culmination of the foresight journey. It builds upon the signals and trends identified, the scenarios developed, and the strategic options evaluated. The playbook takes these insights and translates them into a coherent strategy and action plan.

Here's a step-by-step guide on how to generate the strategic foresight playbook:

1. Define the Vision: a. Create a memorable name for your strategy (e.g., "Autonomous Mobility Network"); b. Write a concise description of the strategy, including key elements and selection criteria; c. Describe the long-term results (7–10 years) for the business and customers; d. Outline the mid-term achievements (3–5 years).

2. Provide Reasons to Believe: a. List qualitative arguments supported by early signals; b. Explain how the strategy differentiates from competitors; c. Describe new user benefits, referencing your scenario; d. Add any additional compelling evidence

3. Address Barriers and Responses: a. Identify potential challenges to the strategy; b. Outline required business responses and necessary changes

4. Amplify Strengths and Reduce Risks: a. Conduct a SWOT analysis based on your situational analysis; b. Explain how to leverage strengths and opportunities; c. Describe methods to mitigate weaknesses and threats

5. Outline Implications for Market-Facing Functions: a. Discuss challenges and opportunities for branding and communication; b. Explain necessary changes to product and service innovation and design

6. Plan for Evolving to Support the Strategy: a. Describe how to build a culture of agile anticipation b. Identify other key operational and organizational success factors

7. Map the Path Forward: a. List immediate next steps to develop urgency b. Prioritize actions based on impact and logical sequence c. Create a visual roadmap of activities and milestones for the next 10+ years

8. Refine and Iterate: a. Use the provided space to draft ideas and make lists b. Continuously refine your roadmap and strategy based on feedback and new insights

Design Futures Layer—Make It Tangible
To translate strategy into lived experience and invite stakeholder sense-making, pair the playbook with design-futures practices:

- Experiential scenarios: Storyboard or "day-in-the-life" vignettes for each strategy variant; include touchpoints, emotions, and breakdowns.

- Diegetic prototypes/artifacts from the future: Mock app screens, curb-use signage, fleet UI, receipts, safety notices, service T&Cs—dated and branded to feel real.

- Future personas and contexts: Riders, drivers, city planners in Sunbelt metros; accessibility and heat-risk personas; equity lenses.

- Location sheets: Future curb layouts, charging depots, micro-mobility hubs (2028/2032 snapshots).

- Experience walk-throughs: Facilitated sessions where stakeholders interact with the artifacts and surface risks, ethics, and policy trade-offs.

- Evidence back-links: Each artifact captions the signals/drivers it instantiates and the assumptions it tests.

Outputs: An artifact catalogue, gallery deck, and facilitation guide aligned to the selected strategy and its "no-regret/option/guardrail" portfolio.

Worked Example: See the RideNow case above for a compact instance of this playbook (strategy, evidence, and the Design Futures layer of scenarios, artifacts, personas, and locations).

Apply to Your Context: Adapt the steps to your organization's goals and constraints, pairing each strategic choice with tangible artifacts for stakeholder testing and iteration.

CHAPTER 9   SYNTHESIZING FUTURES PRACTICE

9.3 Case Study: The Future of Roads and Highways

As we have previously emphasized, we cannot predict the future, but we can imagine what can emerge from all the complexity of relationships in the world between nature, humans, objects, and any other elements outside of this intricate web. That's why the basis of futures thinking is about opening the eyes to everything happening around and the potential paths they might be taking. With all this information including history, we can extrapolate and imagine not only one potential future but multiple ones, that's why it's called "Futures" Thinking.

To frame multiple futures, we use the Futures Cone, popularized by Joseph Voros (orig. c.2003; most recent version 2017), version shown in Figure 6-4. The cone depicts how moving from the present, the space of the possible, plausible, probable, and preferable futures expands over time.

Speculative Design

When studying Dunne and Raby, the authors of *Speculative Everything*, we can see how they have done a masterful work not only in terms of a progressive view about critic design but also its impact in society. They have created artifacts like the "Digicars," based on the premise of the "digitarians" generations. The "Digicars" are electric, autonomous vehicles designed to meet personal needs and optimize roads efficiency that illustrate the technocrat society and its influence in the economy and its artifacts.

Choosing a Focal Topic to Begin the Exploration

1. Determine the focus of the exploration: This can be based on the particular area of interest or particular need in the industry you are located in or curious about. In this example, we are going to work with the Future of Transportation.

CHAPTER 9   SYNTHESIZING FUTURES PRACTICE

2. Explore signals of change: Detecting the clues or trends of the changes that are happening that could be early indications of important changes toward the future.

The process begins with formulating the question of interest. For example, in this case study, the question is "What would be the drivers of roads and highways in 2040?"

This question depends on the particular interest of study or a need in a specific field either for social exploration purposes or strategic planning purposes. Then we might move onto exploring our surroundings with day to day observations about the environment.

For example, where I live, weeks ago without even searching for it, I was waiting in line to get an ice cream in a local park and the people ahead of me were talking about the local city creating local events about the future of transportation in 2050. They were particularly commenting on the use of biking paths and how they might go about this since the winters here in Winnipeg are harsh, sometimes with temperatures of -40C.

When we start to gather info about the environment, we can see weak signals about our topic under study and begin to dig deeper to uncover more detailed insights on how these same issues are being dealt with or considered in other latitudes around the world.

Many years ago, I don't recall seeing so many people riding bikes to work even where I live, which is very cold in the winter. Now, more and more I have started to observe the use of other transportation vehicles, for example, the scooter. This is an example of early signs that might point toward futures where other vehicles will emerge, and its infrastructure along with them.

CHAPTER 9   SYNTHESIZING FUTURES PRACTICE

My second choice of insight gathering is through specialized sites, where we can begin to monitor or even get those trend alerts in our emails. An example of these sites is thefuturescentre.org. Also blogs, leading industry actors or leading companies, news, and the media in general are outlets to explore those signals.

Now, let's discuss the shift between the manual work performed here in comparison with using AI specialized or AI general commercial platforms available. Before we learn how to leverage these tools, we need to make sure that we are aware of the responsibility of its use and risks, which as developed previously, entails that we monitor its outputs to make sure the information is trustable and that it follows ethical standards.

The process to gather signals info using AI can be done in different ways, and here we are about to study three methods.

The first one is to train your own model to be used for this purpose. This is an option for the ones who are advanced with the use of machine learning models and know how to customize or use APIs for models already in the market with trained databases.

A place to begin learning about this is the service offered by Google, called Vertex AI, where you can find different models and even tutorials to follow to leverage a variety of models in the market for use.

The second one is to use direct platforms specialized in AI insights for the process of horizon scanning powered by AI, which means that their system will use an algorithm to search in their databases for that particular topic and will display the particular latest signals categorized in different tags like emerging, peaking, or declining. You can search for other platforms like shapingtomorrow.com where you can search for emerging trends and begin to build your signals.

The third method is to use a current commercial LLM (e.g., Perplexity, Gemini, ChatGPT, Claude) with web/retrieval and visualization features. Choose any up-to-date release available to you and log the tool, model name, and date used in your project notes.

CHAPTER 9   SYNTHESIZING FUTURES PRACTICE

Versioning note: AI tools change rapidly; to keep this guidance durable, we avoid listing specific builds. For reproducibility, record tool, model variant, and access date (e.g., "Claude—Artifacts workspace, accessed May 2025"). If your organization requires, include exact version/build numbers in an internal log.

In this case, I will be using the latest version available of Anthropic at the time of writing this book (2025), since it has great capabilities of using a tool called "Artifacts" that allow you to create comprehensive dynamic visuals based on the research that you are performing. Since these tools advance so fast, years forward more and more of these kinds of models will be available in the market even making it easier and easier to customize our research.

In our example, after defining our question, we might use the most recent model and input a prompt like this keeping in mind that some initial research would be good to make sure we can determine how trustable the information can be.

"What are the best signals in futures design or strategic foresight about the future of roads and highways classified according with impact, likelihood, timeframe, maturity and priority with real information on the real signals happening right now?"

This process was tested with the manual method and using the Pro and latest versions of Perplexity, Gemini, ChatGPT, and Claude from Anthropic. After this comparison, it was possible to conclude after checking the sources that the information was legitimate and the best-performing model for this purpose was Claude from Anthropic. The evaluation resulted in the following outcome.

CHAPTER 9   SYNTHESIZING FUTURES PRACTICE

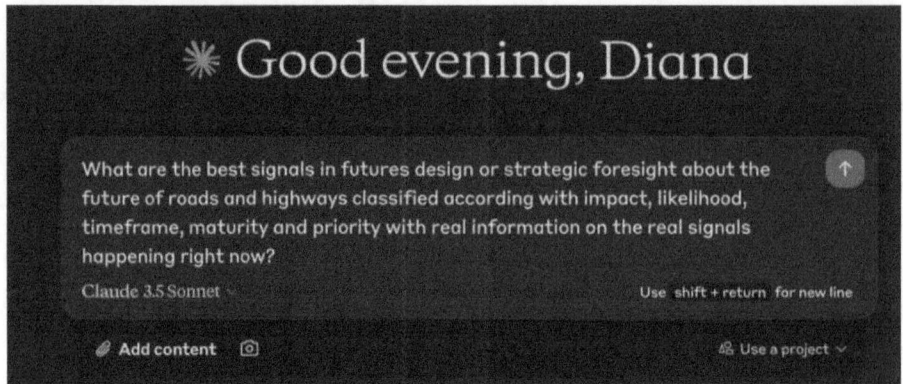

Figure 9-5. *Using Claude to Gather Initial Insights*

Once we have gathered some initial information, supported with all the other sources discussed previously, such as news, signals from specialized sources, direct observations in the market, and intentional research from other additional sources, we might come up with an array of data that we can then upload back into Claude to help with the analysis and classification by asking it using a prompt, which might look like: "Please classify the presented information by Impact, Likelihood, Timeframe, Maturity, and Priority…". Here you can add as many details to your prompt as considered necessary. Then the next step consists of reviewing the information and insights based on the manually captured data to ensure it is reflecting the trends observed.

No-code accuracy workflow (LLM chat + spreadsheet)
Suggested prompt (copy/paste):

You are classifying short items into STEEP+V categories. Pick at most two labels from: Social (S), Technological (T), Economic (Ec), Environmental (En), Political/Policy (P), Values/Ethics (V), and UNSURE:Disambiguation: Laws/regulation ➤ P (even if about tech). Cultural belief ➤ V (not S, unless measured behavior). Firm costs/revenue ➤ Ec. R&D funding policy ➤ P+T.

CHAPTER 9   SYNTHESIZING FUTURES PRACTICE

For each item, return a row in a table with columns:Item ID | Label(s) | Confidence (0–100) | One-sentence justification quoting key phrase(s).

If confidence <70, use UNSURE and say why.

Here are 10 examples to learn the style:

"City council passes EV-only zone" ➤ P (regulation)...

"Breakthrough solid-state battery" ➤ T...

"Riders switch to monthly transit passes" ➤ Ec...

"Heat waves disrupt bus service" ➤ En...

"Gen Z prefers access over ownership" ➤ S...

"Privacy backlash against ride data sharing" ➤ V...

"Tax credits for charging networks" ➤ P+Ec...

"AI dispatch slashes wait times" ➤ T...

"Unionization push among drivers" ➤ P+S...

"Corporate earnings miss on fuel costs" ➤ Ec...

Now classify the following items and return a markdown table.

Paste your items (a bullet list works), and the model will return a table you can copy straight into Excel/Sheets.

Spot-check

Manually review 10–20% of rows.

If you see recurring mistakes (e.g., P vs. V), add a one-line rule to the prompt and rerun.

CHAPTER 9   SYNTHESIZING FUTURES PRACTICE

Signal	Impact	Likelihood	Timeframe	Maturity	Priority
Smart Roads and Infrastructure	High	High	5-10 years	Developing	High
Electric Vehicle (EV) Charging Infrastructure	High	Very High	0-5 years	Growing	Very High
Autonomous Vehicle (AV) Ready Roads	High	Medium	10-15 years	Emerging	Medium
Green Infrastructure and Materials	Medium	High	0-5 years	Growing	High
Dynamic Road Pricing	Medium	Medium	5-10 years	Developing	Medium
Adaptive Traffic Management Systems	Medium	High	0-5 years	Growing	High
Vehicle-to-Everything (V2X) Communication	High	Medium	5-10 years	Emerging	High
Multi-Modal Integration	Medium	High	0-5 years	Growing	High
Weather-Responsive Road Technologies	Medium	Medium	5-10 years	Emerging	Medium
Urban Air Mobility Integration	High	Low	15-20 years	Conceptual	Low

Figure 9-6. *Using Claude to Organize the Data and Classify It*

To be able to gather these signals in a structured way, we might use signals cards so that we can come back to check its contents in an organized way. So we can summarize the finding with a small paragraph and with a visual representation of it. If you have checked my work, you might already know that I like to use concept creation tools to deliver and represent visual representations. In this case, I use Midjourney to generate an image for this summary.

To generate this image, I use Midjourney on the website or Discord with a prompt that can describe each signal. For example, I might start by creating a prompt like: "Electric vehicle and EV charging infrastructure" by which I would get a result such as the following below on image Figure 9-7.

CHAPTER 9 SYNTHESIZING FUTURES PRACTICE

Figure 9-7. *Using Midjourney to Generate Initial Visuals*

This process requires iterations to be able to find an image suitable for our study. In this case, we came up with this one:

Figure 9-8. *Example of Image Generated by Midjourney*

Then using Figma, we have moved onto creating a card that describes the summary of the signal found with the respective link for easy access. This process is repeated with every relevant signal found. And finally the first card looks like this (I am using here Figma to create the card framework):

CHAPTER 9   SYNTHESIZING FUTURES PRACTICE

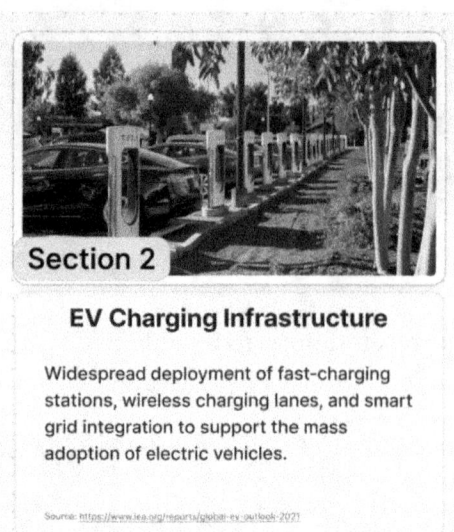

Figure 9-9. Signal Card Created with Figma

Following the same process with each other cards, we might arrive to a set of signals like this:

Figure 9-10. Collection of Signals Ready

CHAPTER 9   SYNTHESIZING FUTURES PRACTICE

And then we would move to prepare the STEEP+V analysis which also usually has a couple different methods for execution. In this case, we will create it by categorizing each of the signals in each section. As discussed before, STEEP+V stands for S: Society, T: Technology, E: Environment, E:Economy, P:Political, and V:Values. After this categorization, this is the result:

Figure 9-11. *Signals Categorized for STEEP+V*

Then in this next step, we proceed to further analyze each category providing additional insights on the implications of each section for the analysis.

STEEP+V Analysis: Future of Roads and Highways

Social

CHAPTER 9 SYNTHESIZING FUTURES PRACTICE

1. ***Changing Mobility Patterns***
   - *Shift toward shared mobility and Mobility-as-a-Service (MaaS)*
   - *Increased acceptance of autonomous vehicles*
   - *Growing preference for multimodal transportation options*

2. ***Urbanization and Smart Cities***
   - *Integration of road systems with smart city infrastructure*
   - *Adaptation of roads to support higher population densities*
   - *Emphasis on walkability and livable urban spaces*

3. ***Aging Population***
   - *Design of roads to accommodate older drivers and pedestrians*
   - *Integration of assistive technologies in road infrastructure*
   - *Increased focus on accessibility in transportation design*

4. ***Public Health Considerations***
   - *Roads designed to promote active transportation (walking, cycling)*
   - *Integration of air quality monitoring and mitigation measures*
   - *Noise reduction technologies in road construction*

CHAPTER 9   SYNTHESIZING FUTURES PRACTICE

5. ***Safety Expectations***
   - *Zero-fatality vision driving road safety innovations*
   - *Increased public expectation for real-time safety information*
   - *Growing acceptance of technology-driven safety measures*

Technological

1. ***Smart Roads and Infrastructure***
   - *Integration of IoT sensors for real-time monitoring*
   - *AI-powered traffic management systems*
   - *Self-healing materials for road maintenance*

2. ***Electric Vehicle (EV) Charging Infrastructure***
   - *Wireless charging lanes*
   - *Ultra-fast charging stations along highways*
   - *Smart grid integration for load balancing*

3. ***Autonomous Vehicle (AV)-Ready Roads***
   - *V2X communication systems*
   - *High-precision lane markings and signage for machine vision*
   - *Dedicated AV lanes with specialized infrastructure*

4. ***Advanced Materials***
   - *Photocatalytic surfaces for pollution reduction*
   - *Nanotechnology in road construction for increased durability*
   - *Temperature-responsive materials for weather adaptation*

5. ***5G and Beyond***
   - *Ultra-low latency communication for real-time traffic management*
   - *Edge computing nodes along highways for data processing*
   - *Improved connectivity for in-vehicle services and entertainment*

6. ***Artificial Intelligence and Big Data***
   - *Predictive maintenance using AI and sensor data*
   - *Machine learning for optimizing traffic flow*
   - *Big data analytics for long-term infrastructure planning*

7. ***Augmented Reality (AR) in Road Use***
   - *AR-enhanced navigation systems integrated with road infrastructure*
   - *Real-time hazard highlighting for drivers and pedestrians*
   - *AR for maintenance and construction workers*

CHAPTER 9   SYNTHESIZING FUTURES PRACTICE

Economic

1. **Dynamic Road Pricing**
   - *Real-time congestion charging*
   - *Usage-based taxation models*
   - *Incentive systems for off-peak travel*

2. **Infrastructure Funding Models**
   - *Public-private partnerships for road development*
   - *Blockchain-based micropayments for road use*
   - *Carbon credit systems tied to green infrastructure*

3. **Economic Impact of Autonomous Vehicles**
   - *Changes in logistics and supply chain efficiencies*
   - *Reduction in accident-related costs*
   - *Shift in employment patterns (e.g., professional drivers)*

4. **Maintenance and Lifecycle Costs**
   - *Predictive maintenance reducing overall costs*
   - *Longer-lasting materials decreasing frequency of repairs*
   - *Integration of revenue-generating technologies (e.g., solar roads)*

CHAPTER 9   SYNTHESIZING FUTURES PRACTICE

5. ***New Business Models***

    - *Road-side service stations evolving for EV and autonomous era*
    - *Data monetization from smart road infrastructure*
    - *Emergence of specialized road construction and maintenance firms*

6. ***Economic Resilience***

    - *Design of road systems to better withstand and recover from economic shocks*
    - *Integration with local economic development plans*
    - *Flexible infrastructure adaptable to changing economic needs*

Environmental

1. ***Green Infrastructure and Materials***

    - *Use of recycled and sustainable materials in road construction*
    - *Integration of carbon-capturing technologies*
    - *Permeable pavements for improved water management*

2. ***Renewable Energy Integration***

    - *Solar panel roads generating electricity*
    - *Wind turbines integrated into highway infrastructure*
    - *Kinetic energy harvesting from vehicle movements*

CHAPTER 9   SYNTHESIZING FUTURES PRACTICE

3. ***Biodiversity Protection***

    - *Wildlife corridors and green bridges over highways*
    - *Habitat-friendly lighting along roads*
    - *Use of native plants in roadside landscaping*

4. ***Climate Change Adaptation***

    - *Resilient road design for extreme weather events*
    - *Coastal road protection against rising sea levels*
    - *Heat-resistant materials for increasing temperatures*

5. ***Emissions Reduction***

    - *Support for zero-emission vehicles (electric, hydrogen)*
    - *Traffic optimization to reduce idle times and emissions*
    - *Integration of air purification technologies in road infrastructure*

6. ***Noise Pollution Mitigation***

    - *Sound-absorbing road surfaces*
    - *Noise barriers with integrated renewable energy generation*
    - *Active noise control systems along highways*

CHAPTER 9   SYNTHESIZING FUTURES PRACTICE

Political

1. **Regulatory Frameworks**
   - *Development of laws for autonomous vehicle operation*
   - *Standardization of smart road technologies*
   - *Privacy and data protection policies for connected infrastructure*

2. **Cross-Border Coordination**
   - *International standards for connected and autonomous vehicles*
   - *Collaborative projects for cross-border smart highways*
   - *Harmonization of road pricing and taxation across regions*

3. **Urban Planning Policies**
   - *Integration of transportation planning with urban development*
   - *Policies promoting multimodal transportation options*
   - *Zoning laws adapted for changing mobility patterns*

4. **Public-Private Partnerships**
   - *Government collaboration with tech companies for infrastructure development*
   - *Privatization vs. public ownership debates for road infrastructure*
   - *Balancing public good with private sector innovation*

CHAPTER 9   SYNTHESIZING FUTURES PRACTICE

5. ***Security and Cybersecurity***

    - *Protection of critical road infrastructure from cyber threats*
    - *Anti-terrorism design in key transportation hubs*
    - *Resilience planning for potential disruptions*

6. ***Ethical Considerations in Autonomous Systems***

    - *Development of ethical frameworks for AV decision-making*
    - *Liability laws for accidents involving autonomous vehicles*
    - *Balancing efficiency with fairness in traffic management systems*

Values

1. ***Sustainability***

    - *Growing emphasis on environmentally friendly transportation*
    - *Value placed on reducing carbon footprint of road infrastructure*
    - *Shift toward circular economy principles in road construction*

2. ***Accessibility and Inclusivity***

    - *Universal design principles in road and transport systems*
    - *Ensuring equitable access to mobility options*
    - *Addressing the digital divide in smart transportation systems*

3. ***Privacy vs. Connectivity***
    - *Balancing benefits of data-driven systems with personal privacy*
    - *Transparency in data collection and use from road infrastructure*
    - *Right to anonymity in public spaces vs. safety and efficiency gains*

4. ***Safety and Trust***
    - *Building public trust in autonomous transportation systems*
    - *Valuing human life and safety above efficiency in system design*
    - *Ethical considerations in programming safety algorithms*

5. ***Work–Life Balance***
    - *Road systems supporting flexible working patterns*
    - *Valuing reduced commute times and improved quality of life*
    - *Integration of leisure and community spaces in transport hubs*

6. ***Community and Place-Making***
    - *Roads as public spaces fostering community interaction*
    - *Preservation of local character in infrastructure development*
    - *Balancing global connectivity with local identity*

CHAPTER 9   SYNTHESIZING FUTURES PRACTICE

7. ***Intergenerational Equity***

    - *Long-term thinking in infrastructure development*

    - *Consideration of future generations in resource use and planning*

    - *Adaptability of road systems to changing societal needs*

Next, with this analysis in mind, we proceed to find patterns or trends in our identified signals. After analyzing them, we can see three prominent patterns: Intelligent Transportation Systems, Sustainable and Resilient Infrastructure, and Future Mobility Integration.

Then we will move to create images of the future in a creative way. This is where our Futures Wheel comes to place.

The futures wheel is used to analyze and organize the implications of the changes identified or concrete events. It is in essence a visual map of the implications of certain chain of events toward the future.

Let's begin with the most relevant topic for this analysis. In this case, let's work with Intelligent Transportation Systems. What would be its direct consequences?

Here we are going to categorize them in short term (first order), medium term (second order) and long term (third order). The key here is that the futures wheel might indicate the future could look like the third-order consequences.

First-Order Consequences focus on immediate effects such as improved traffic flow, enhanced safety, optimized energy consumption, increased data collection, new infrastructure needs, and changes in driver behavior.

Second-Order Consequences delve into broader impacts, including urban planning transformation, economic shifts, insurance industry disruption, changes in public transportation, environmental effects, and cybersecurity challenges.

CHAPTER 9   SYNTHESIZING FUTURES PRACTICE

Third-Order Consequences explore long-term, systemic changes such as societal shifts, healthcare impacts, legal and ethical considerations, educational system adaptations, psychological effects, and impacts on global competitiveness.

Futures Wheel: Intelligent Transportation Systems
First-Order Consequences for Intelligent Transportation Systems

1. Improved Traffic Flow
    - Reduced congestion in urban areas
    - Shorter travel times for commuters
    - More predictable arrival times for logistics and public transport
2. Enhanced Road Safety
    - Decreased number of accidents
    - Quicker response times for emergency services
    - Real-time hazard warnings for drivers
3. Optimized Energy Consumption
    - Reduced fuel consumption due to smoother traffic flow
    - Lower emissions from vehicles
    - More efficient use of electric vehicle batteries
4. Increased Data Generation and Collection
    - Real-time insights into traffic patterns
    - Detailed usage statistics for infrastructure planning
    - Potential privacy concerns for road users

CHAPTER 9   SYNTHESIZING FUTURES PRACTICE

5. New Infrastructure Requirements
   - Need for sensor networks along roads
   - Upgrade of existing road signs and signals
   - Installation of roadside computing units
6. Changes in Driver Behavior
   - Increased reliance on in-vehicle information systems
   - Potential for distraction from multiple information sources
   - Adaptation to new traffic management strategies

Second-Order Consequences

1. Urban Planning Transformation
   - Redesign of city layouts to maximize intelligent traffic flow
   - Repurposing of some urban spaces due to reduced parking needs
   - Integration of transportation hubs with smart city initiatives
2. Economic Impacts
   - New job creation in ITS maintenance and development
   - Potential job losses in traditional traffic management roles
   - Emergence of new business models based on traffic data

3. Insurance Industry Disruption
   - Shift in liability determinations for accidents
   - New insurance products based on real-time driving data
   - Potential reduction in insurance premiums due to increased safety
4. Changes in Public Transportation
   - More efficient and reliable public transit systems
   - Increased integration between private and public transportation
   - Potential for new forms of on-demand, AI-driven public transit
5. Environmental Effects
   - Improved air quality in urban areas
   - Reduced noise pollution from smoother traffic flow
   - Potential increase in electronic waste from ITS components
6. Cybersecurity Challenges
   - Increased vulnerability to cyber-attacks on transportation infrastructure
   - Need for robust security measures to protect connected systems
   - Potential for large-scale disruptions if systems are compromised

CHAPTER 9   SYNTHESIZING FUTURES PRACTICE

Third-Order Consequences

1. Societal Shifts

    - Changes in work patterns due to more predictable commutes
    - Potential for increased suburban sprawl enabled by efficient transportation
    - Evolution of social norms around vehicle ownership and use

2. Healthcare Impacts

    - Reduction in respiratory illnesses due to decreased pollution
    - Fewer hospitalizations from traffic accidents
    - Potential increase in sedentary lifestyles due to ultra-efficient transport

3. Legal and Ethical Considerations

    - New laws and regulations governing intelligent transportation systems
    - Ethical debates around data privacy and algorithmic decision-making
    - Potential for social inequity based on access to smart transportation

4. Educational System Adaptations

    - New curricula focused on ITS engineering and management
    - Changes in driver education programs
    - Increased emphasis on data literacy and cybersecurity in general education

5. Psychological Effects
    - Reduced stress from improved traffic conditions
    - Potential anxiety from increased technological dependence
    - Changes in perception of control and autonomy in transportation
6. Global Competitiveness
    - Cities' attractiveness tied to the efficiency of their transportation systems
    - National economic advantages based on early adoption of ITS
    - New parameters for quality of life rankings incorporating transportation efficiency

CHAPTER 9   SYNTHESIZING FUTURES PRACTICE

An initial visual might look like this:

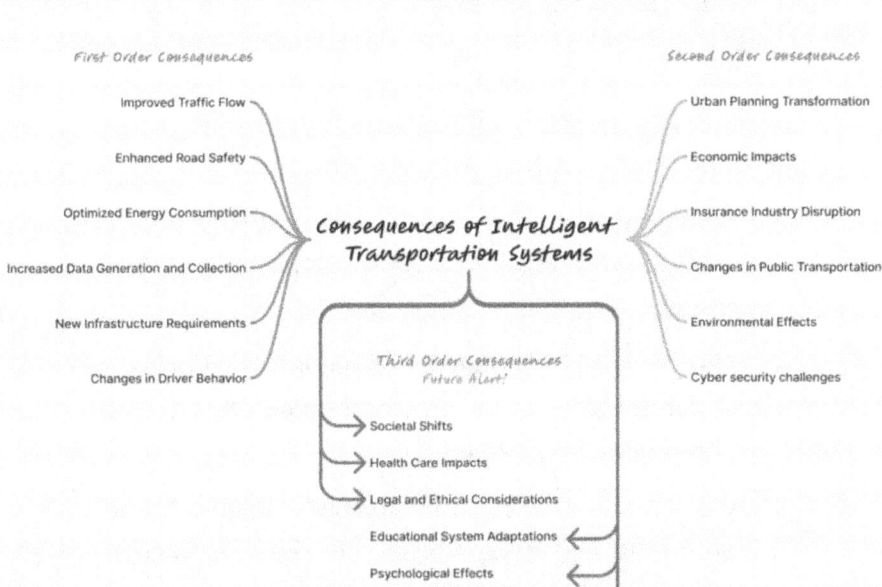

Figure 9-12. Futures Wheel for Intelligent Transportation Systems

Future Vision: The World Shaped by Intelligent Transportation Systems

1. **Redefined Urban Landscapes**

   In a future dominated by Intelligent Transportation Systems (ITS), cities will likely undergo significant transformations:

   - Cities may become more spread out due to efficient transportation, with "smart suburbs" emerging as attractive living options.

   - Urban centers could see a reduction in surface parking, replaced by green spaces, pedestrian zones, or community areas.

CHAPTER 9   SYNTHESIZING FUTURES PRACTICE

- We might witness the rise of "15-minute cities," where all essential services are accessible within a short, efficient journey.

2. **Shift in Work-Life Patterns**

The predictability and efficiency of travel could reshape how and where people work:

- Remote work might become even more prevalent, with people choosing to live further from city centers.

- Flexible working hours could become the norm, as commute times become more reliable and potentially productive.

- We may see a rise in "third spaces"—neither home nor traditional office—as people take advantage of their ability to work from anywhere.

3. **Health and Well-Being Implications**

The health landscape could shift significantly:

- Improved air quality in urban areas may lead to a decrease in respiratory illnesses.

- There might be a dramatic reduction in traffic-related injuries and deaths.

- However, we may face new health challenges related to increasingly sedentary lifestyles facilitated by ultra-efficient, door-to-door transportation.

CHAPTER 9   SYNTHESIZING FUTURES PRACTICE

4. **Evolving Social Norms and Behaviors**

   Transportation efficiency could lead to new social dynamics:

   - The concept of vehicle ownership might shift dramatically, with more people opting for shared, on-demand transportation services.

   - Social interactions could change, with people potentially having more free time due to reduced commutes.

   - There might be a shift in status symbols away from car ownership to other forms of consumption or experiences.

5. **New Educational and Career Landscapes**

   Education and career paths will likely adapt to this new transportation reality:

   - New fields of study and career paths in ITS engineering, data science, and system management will emerge.

   - There will likely be a greater emphasis on continuous learning and re-skilling, as traditional transportation-related jobs evolve or disappear.

   - Education itself might become more decentralized, with students able to access schools and universities further from home.

6. **Legal and Ethical Frameworks**

   A new legal and ethical landscape will need to emerge:

   - We may see the development of a comprehensive "Digital Transportation Rights" framework, addressing issues of privacy, access, and fairness.

   - There could be ongoing debates about the ethics of algorithmic decision-making in traffic management and its impact on different socio-economic groups.

   - New forms of "transportation citizenship" might emerge, with rights and responsibilities tied to one's participation in the intelligent transportation ecosystem.

7. **Psychological Adaptation**

   People's relationship with transportation and technology will evolve:

   - While stress related to traffic and commuting may decrease, new forms of anxiety related to technological dependence might emerge.

   - The concept of "freedom of movement" might be redefined, balancing efficiency with the human desire for control and spontaneity.

   - There may be a growing emphasis on "digital detox" and "analog transportation" experiences as a counterbalance to the highly connected transportation environment.

CHAPTER 9   SYNTHESIZING FUTURES PRACTICE

8. **Global and Economic Shifts**

   On a larger scale, ITS could reshape global dynamics:

   - Cities and nations with advanced ITS might gain significant economic advantages, potentially reshaping global competitiveness.

   - Quality of life rankings may be increasingly tied to the efficiency and sustainability of transportation systems.

   - We might see new forms of "transportation diplomacy," with nations collaborating (or competing) on large-scale, cross-border intelligent transportation projects.

Now, let's move onto the next group of analysis obtained from the signals categories: Sustainable and Resilience Infrastructure.

Futures Wheel: Sustainable and Resilient Infrastructure
First-Order Consequences

1. Reduced Environmental Impact

   - Lower carbon emissions from road construction and maintenance

   - Decreased use of nonrenewable resources in road materials

   - Minimized disruption to local ecosystems

CHAPTER 9   SYNTHESIZING FUTURES PRACTICE

2. Improved Climate Resilience

    - Roads better able to withstand extreme weather events
    - Reduced damage from heat, cold, flooding, and storms
    - Fewer road closures due to weather-related incidents

3. Enhanced Water Management

    - Improved stormwater runoff handling
    - Reduced water pollution from road surfaces
    - Better groundwater recharge in urban areas

4. Integration of Renewable Energy

    - Solar roads generating electricity
    - Kinetic energy harvesting from vehicle movement
    - Wind energy capture along highways

5. New Material Technologies

    - Development of self-healing road surfaces
    - Use of recycled and upcycled materials in construction
    - Implementation of pollution-absorbing materials

6. Changes in Construction Practices

    - Adoption of new techniques for sustainable road building
    - Increased focus on lifecycle assessment in project planning
    - Shift toward modular and easily repairable road designs

CHAPTER 9 SYNTHESIZING FUTURES PRACTICE

Second-Order Consequences

1. Economic Shifts

    - Growth in green construction and materials industries

    - Potential higher initial costs but lower long-term maintenance expenses

    - New job creation in sustainable infrastructure sectors

2. Policy and Regulation Changes

    - Implementation of stricter environmental standards for road construction

    - Development of new certifications for sustainable infrastructure

    - Changes in government funding priorities for road projects

3. Urban Planning Transformations

    - Integration of green corridors along roadways

    - Increased focus on multiuse infrastructure designs

    - Reimagining of urban spaces with sustainable roads as a central element

4. Public Health Improvements

    - Reduction in air and water pollution–related illnesses

    - Decreased urban heat island effect

    - Potential increase in green spaces along roadways

CHAPTER 9 SYNTHESIZING FUTURES PRACTICE

5. Changes in Transportation Behavior
    - Encouragement of eco-friendly transportation modes
    - Potential for new types of vehicles designed for sustainable roads
    - Shift in public perception of road infrastructure
6. Resilience to Climate Change
    - Reduced economic losses from climate-related infrastructure damage
    - Improved emergency response capabilities during extreme weather
    - Enhanced food security through more reliable transportation networks

Third-Order Consequences

1. Shift in Global Resource Demands
    - Decreased demand for traditional road-building materials
    - Increased demand for rare earth elements used in advanced road technologies
    - Potential geopolitical shifts based on new resource importances
2. Evolution of Urban–Rural Dynamics
    - Improved connectivity between urban and rural areas
    - Potential for more distributed population due to resilient infrastructure
    - Changes in land use patterns and property values

CHAPTER 9   SYNTHESIZING FUTURES PRACTICE

3. Impacts on Biodiversity
   - Creation of new wildlife corridors along green roads
   - Potential recovery of some ecosystems due to reduced pollution
   - New challenges in balancing infrastructure needs with habitat preservation

4. Cultural and Social Shifts
   - Changes in societal values toward infrastructure and environment
   - Potential for new forms of community engagement with road spaces
   - Shift in national identity and pride related to sustainable infrastructure

5. Educational System Adaptations
   - New academic disciplines focusing on sustainable infrastructure
   - Changes in engineering and urban planning curricula
   - Increased emphasis on interdisciplinary approaches to infrastructure

6. Global Competitiveness Realignment
   - Nations' standing influenced by their sustainable infrastructure capabilities
   - New parameters for measuring national development and progress
   - Potential for new forms of international collaboration on infrastructure projects

7. Insurance and Finance Industry Transformations
    - New models for assessing infrastructure risk and resilience
    - Development of financial products tied to sustainable road projects
    - Shifts in government bond markets related to infrastructure funding
8. Psychological and Health Effects
    - Reduced stress from improved environmental conditions
    - Changes in people's relationship with their built environment
    - Potential improvements in overall quality of life and well-being

CHAPTER 9 SYNTHESIZING FUTURES PRACTICE

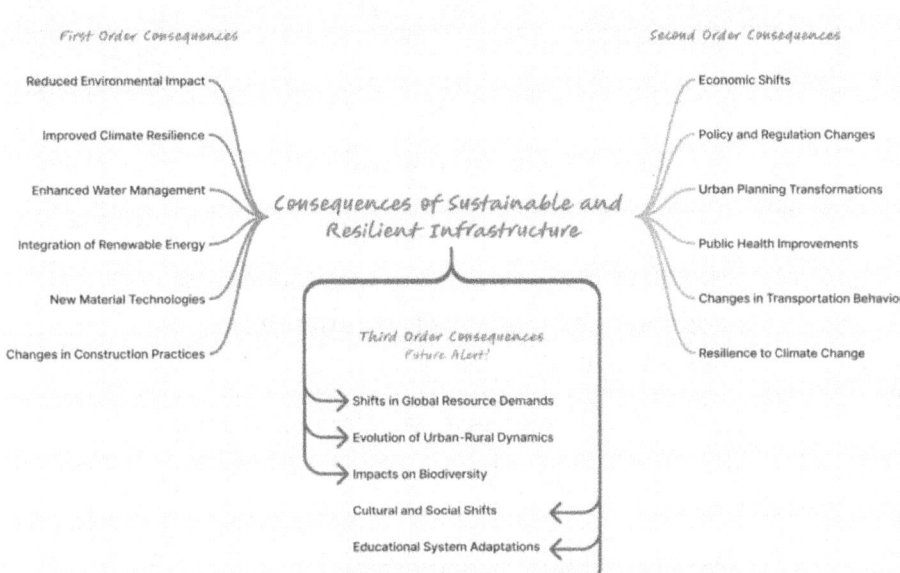

Figure 9-13. *Futures Wheel for Sustainable and Resilient Infrastructure*

Future Vision: World Shaped by Sustainable and Resilient Infrastructure

1. **Reimagined Urban-Rural Dynamics**

   In a future dominated by sustainable and resilient infrastructure

   - The distinction between urban and rural areas may blur, with green corridors connecting cities and countryside.

   - We might see the emergence of "eco-towns"—self-sufficient communities built around sustainable infrastructure networks.

- Rural areas could become more accessible and attractive for living, potentially leading to a more distributed population.

2. **New Relationship with Nature**

   The integration of infrastructure with natural systems could lead to

   - Cities that function more like ecosystems, with roads and buildings actively contributing to biodiversity.
   - A cultural shift toward viewing human-made structures as part of nature rather than separate from it.
   - Increased public engagement with local environments, fostering a stronger conservation ethic.

3. **Resilient Societies**

   The emphasis on resilient infrastructure might result in

   - Communities that can better withstand and quickly recover from extreme weather events and natural disasters.
   - Reduced economic losses from climate-related disruptions, leading to more stable local economies.
   - A sense of security and optimism about the future, even in the face of climate change.

CHAPTER 9   SYNTHESIZING FUTURES PRACTICE

4. **Shifted Global Power Dynamics**

    The focus on sustainable infrastructure could reshape global politics:

    - Nations might be ranked and gain influence based on their sustainability metrics and infrastructure resilience.

    - New alliances could form around sustainable technology and resource sharing.

    - There might be a shift in economic power toward countries rich in materials crucial for sustainable infrastructure.

5. **Transformed Education and Career Landscapes**

    The needs of sustainable infrastructure will likely reshape learning and work:

    - We may see the rise of new academic fields integrating ecology, engineering, and urban planning.

    - There could be a surge in demand for professionals skilled in sustainable design, resilient systems, and environmental restoration.

    - Continuous learning might become the norm as infrastructure systems evolve and improve.

6. **Innovative Financial Systems**

    The economy might adapt to support long-term sustainability:

    - New financial products could emerge, such as long-term infrastructure bonds or resilience insurance.

- We might see a shift toward valuing long-term sustainability over short-term gains in investment strategies.

- The concept of national wealth might expand to include measures of environmental health and infrastructure resilience.

7. **Enhanced Public Health and Well-Being**

Sustainable infrastructure could lead to significant health improvements:

- Reduced pollution levels could result in lower rates of respiratory and cardiovascular diseases.

- Increased green spaces and better integration with nature might improve mental health outcomes.

- Active transportation infrastructure could combat sedentary lifestyles, reducing obesity and related health issues.

8. **Evolving Cultural Norms and Values**

Society's relationship with infrastructure and the environment may fundamentally change:

- There might be a cultural shift toward valuing durability, repairability, and sustainability in all aspects of life.

- New forms of recreation and social interaction could emerge around green infrastructure spaces.

- The aesthetic of cities might change dramatically, with a new appreciation for the beauty of functional, sustainable design.

9. **Redefined Concept of Progress**

   How we measure societal advancement might evolve:

   - Traditional economic growth metrics might be replaced or supplemented by measures of sustainability and resilience.

   - The idea of a "developed" nation could be redefined to prioritize environmental stewardship and infrastructure adaptability.

   - There might be a global shift toward prioritizing quality of life and environmental health over consumption-based measures of success.

And finally, let's study our third strongest trend: Future Mobility Integration.

Futures Wheel: Future Mobility Integration

First-Order Consequences

1. Autonomous Vehicle Dominance

   - Widespread adoption of self-driving cars

   - Reduction in human-driven vehicles

   - Changes in road design to accommodate AV communication needs

2. Multimodal Transportation Hubs

   - Integration of various transport modes (e.g., cars, bikes, public transit, air taxis)

   - Seamless transfer between different transportation methods

   - New infrastructure for mode-switching points

3. Personalized Mobility Services
   - Growth of on-demand transportation options
   - Decline in personal vehicle ownership
   - Customized travel experiences based on user preferences
4. Changes in Road Usage Patterns
   - More efficient use of road space
   - Potential for narrower lanes due to precise AV navigation
   - Dynamic lane allocation based on real-time demand
5. Urban Air Mobility Integration
   - Introduction of flying taxis and drones for passenger transport
   - New air corridors over existing road networks
   - Vertiports integrated into urban infrastructure
6. Enhanced Accessibility
   - Improved mobility options for elderly and disabled individuals
   - Increased independence for non-drivers
   - More equitable access to transportation across socioeconomic groups

CHAPTER 9 SYNTHESIZING FUTURES PRACTICE

Second-Order Consequences

1. Urban Planning Revolution

    - Redesign of cities to accommodate new mobility patterns
    - Repurposing of parking spaces and gas stations
    - Integration of transportation with residential and commercial spaces

2. Economic Shifts

    - Disruption in traditional automotive and transportation industries
    - New job creation in AV technology, mobility services, and infrastructure management
    - Changes in real estate values based on new transportation patterns

3. Energy Infrastructure Adaptation

    - Widespread electric vehicle charging networks
    - Potential for wireless charging roads
    - Integration of renewable energy sources with transportation systems

4. Data Management Challenges

    - Massive increase in data generation from connected vehicles and infrastructure
    - Need for robust cybersecurity measures
    - Privacy concerns and new regulations for mobility data

CHAPTER 9   SYNTHESIZING FUTURES PRACTICE

5. Changes in Social Interaction
   - New forms of social engagement during travel
   - Potential for mobile offices and entertainment spaces
   - Shift in the concept of "commute time" as productive or leisure time

6. Environmental Impacts
   - Potential reduction in emissions due to efficient routing and electric vehicles
   - Changes in urban heat island effect due to different road usage
   - New considerations for wildlife crossings and ecosystem fragmentation

Third-Order Consequences

1. Redefined Concept of Distance
   - Changes in perception of travel time and distance
   - Potential for increased urban sprawl due to comfortable long-distance travel
   - Shift in work–life balance with new commuting paradigms

2. Healthcare System Adaptation
   - Changes in emergency response and mobile healthcare delivery
   - Potential reduction in traffic-related injuries and deaths
   - New health considerations related to increased sedentary travel

CHAPTER 9 SYNTHESIZING FUTURES PRACTICE

3. Education System Transformation
   - Rise of mobile classrooms and distance learning opportunities
   - Changes in school districting and student transportation
   - New curricula focused on autonomous systems and mobility technologies

4. Legal and Ethical Frameworks
   - Development of new laws governing autonomous and diverse mobility systems
   - Ethical considerations in programming decision-making for AVs
   - New insurance and liability models for multimodal, autonomous transport

5. Psychological Impacts
   - Changes in stress levels associated with commuting
   - Potential loss of driving as a rite of passage or skill
   - New forms of motion sickness or anxiety related to autonomous travel

6. Global Connectivity
   - Seamless international travel with integrated autonomous systems
   - Potential for new forms of global mobility services
   - Changes in immigration and border control processes

CHAPTER 9   SYNTHESIZING FUTURES PRACTICE

7. Cultural Shifts

    - Evolution of car culture and associated social status
    - New forms of entertainment and services designed for autonomous travel
    - Changes in urban–rural dynamics and cultural exchange

8. Economic Globalization Effects

    - Increased efficiency in goods transportation
    - Changes in global supply chains and just-in-time delivery systems
    - Potential for new economic corridors based on autonomous transportation networks

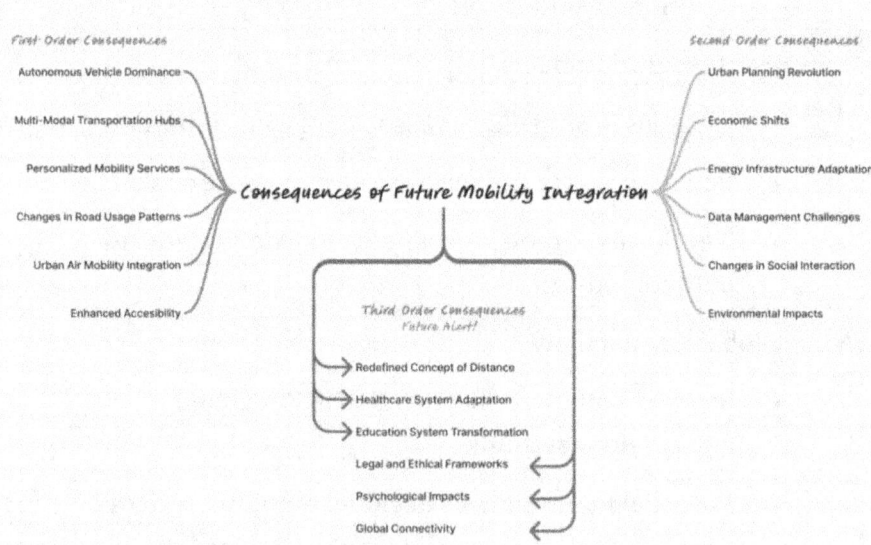

Figure 9-14. Futures Wheel for Future Mobility Integration

CHAPTER 9   SYNTHESIZING FUTURES PRACTICE

Future Vision: The World Shaped by Future Mobility Integration

1. **Redefined Concept of Distance and Living Patterns**

   In a world of diverse and autonomous transportation networks

   - The traditional notion of "commuting distance" may become obsolete, potentially leading to more dispersed living patterns.

   - Urban sprawl could increase as people choose to live further from city centers, knowing they can travel comfortably and productively.

   - We might see the emergence of "super-commuters" who live in one city and work in another, traveling long distances daily in autonomous vehicles.

2. **Transformed Healthcare Delivery**

   The healthcare system may adapt to new mobility patterns:

   - Mobile healthcare units could become prevalent, bringing services directly to patients in autonomous vehicles.

   - Emergency response systems might be revolutionized by autonomous vehicles and urban air mobility, potentially saving more lives.

   - New health challenges related to increased sedentary travel may emerge, prompting innovations in in-vehicle exercise and health monitoring.

3. **Reimagined Education System**

   Education could be fundamentally reshaped:

   - The concept of a local school district might evolve as students can easily travel longer distances for specialized education.

   - Mobile classrooms in autonomous vehicles could bring education to remote or underserved areas.

   - Curricula might shift to include more focus on autonomous systems, AI, and new mobility technologies, preparing students for future careers.

4. **New Legal and Ethical Landscapes**

   Complex legal and ethical frameworks may need to be developed:

   - New laws governing autonomous vehicles and diverse mobility systems could emerge, potentially creating a new field of "mobility law."

   - Ethical guidelines for AI decision-making in transportation scenarios might become a crucial area of philosophical and legal debate.

   - Insurance and liability models could shift dramatically, possibly moving from individual driver insurance to product liability for autonomous vehicle manufacturers.

5. **Psychological and Cultural Shifts**

   The society's relationship with transportation might fundamentally change:

- The stress of driving and traffic could be significantly reduced, potentially improving overall mental health.

- New forms of motion sickness or anxiety related to autonomous or aerial travel might emerge, creating new fields in psychology.

- The cultural significance of getting a driver's license as a rite of passage might diminish, replaced by other milestones related to autonomous mobility.

6. **Enhanced Global Connectivity**

   A new era of global mobility could emerge:

   - International travel might become more seamless, potentially changing immigration processes and cultural exchange.

   - We might see the rise of truly global citizens, easily moving between countries for work, education, or leisure.

   - New forms of global mobility services could arise, offering seamless multicountry transportation packages.

7. **Evolution of Urban–Rural Dynamics**

   The relationship between urban and rural areas could be transformed:

   - The distinction between urban and rural living might blur, with people more freely choosing where to live based on preference rather than proximity to work.

CHAPTER 9   SYNTHESIZING FUTURES PRACTICE

- Rural areas could see revitalization as they become more accessible, potentially reversing urbanization trends.

- New types of communities might emerge along major autonomous transportation routes, similar to how towns historically grew along railways.

8. **Economic Restructuring**

    The economy could shift to accommodate and capitalize on new mobility patterns:

    - Traditional automotive industries might be disrupted, while new industries around autonomous systems and mobility services could flourish.

    - Global supply chains could be revolutionized by autonomous long-haul transportation, potentially reshaping trade patterns.

    - New economic corridors might develop along major autonomous transportation routes, influencing regional development patterns.

Now that we have finished this work, it's time to apply the methodology "What If" based on the four archetypes of Jim Dator. So far what is your first impression about this analysis? Considering that just because we cannot imagine potential futures, doesn't mean that they cannot happen; we can still have important reflections about what outcomes are revealing so far. Are the futures under current study, dystopian, prosperous or undesirable? Based on our early results, these can be categorized in the archetypes created by Jim Dator, who is a futurist specialized in the analysis of these potential futures methodologies. He developed this theory to understand better the behaviors and features of each scenario.

CHAPTER 9 SYNTHESIZING FUTURES PRACTICE

The four archetypes are as follows: Continuation or growth, a world that looks like the one where we live in right now or it tends to trend to become better, associated with economic development. Collapse, which tends toward negative impacts; the third one is Discipline, where the world in which what is considered positive would be upgraded and what would be considered negative would be eliminated. And finally, the fourth one is Transformation, where the trend is totally different to our present conditions, usually emerging from technological developments.

Four Future Scenarios for Transportation and Infrastructure

1. **Continuation or Growth Scenario: "The Hyper-Connected World"**

    What if current trends in technology and mobility continue to accelerate?

    - Smart roads and AI-driven traffic management become ubiquitous, drastically reducing congestion and accidents.

    - Autonomous vehicles dominate, with personal ownership declining in favor of mobility-as-a-service.

    - Urban air mobility becomes commonplace for both personal transport and logistics.

    - Sustainable materials and energy systems are incrementally integrated into existing infrastructure.

    - Global transportation networks become increasingly integrated, facilitating seamless international travel and trade.

237

Implications:

- Increased productivity as commute times are repurposed for work or leisure.
- Continued urban sprawl as distance becomes less of a barrier.
- Gradual improvement in environmental metrics, but potential rebound effects due to increased ease of travel.

2. **Collapse Scenario: "The Fragmented Roads"**

What if economic crises, resource scarcity, or environmental disasters disrupt our transportation systems?

- Maintenance of existing road infrastructure becomes impossible, leading to deteriorating conditions.
- Fuel shortages and high costs make long-distance travel a luxury.
- Autonomous vehicle development stalls due to lack of investment and supporting infrastructure.
- Climate change leads to frequent disruptions of transportation networks due to extreme weather events.
- Local communities become more isolated, relying on local resources and simplified transportation methods.

Implications:

- Resurgence of walking, cycling, and animal-powered transport in many areas.

- Increased importance of local food production and self-sufficiency.

- Potential for new forms of small-scale, resilient transportation solutions to emerge.

3. **Discipline Scenario: "The Regulated Mobility"**

What if strict regulations are imposed to address environmental and social concerns?

- Governments implement stringent carbon quotas for personal travel.

- Private car ownership is heavily restricted in urban areas, replaced by public transit and shared mobility services.

- Massive investment in sustainable infrastructure, including green materials and renewable energy integration in roads.

- Speed limits are lowered and strictly enforced to maximize energy efficiency.

- Urban planning prioritizes compact, walkable cities with extensive bicycle infrastructure.

Implications:

- Significant reduction in carbon emissions from the transportation sector.

CHAPTER 9   SYNTHESIZING FUTURES PRACTICE

- Changes in lifestyle with more emphasis on local community and less long-distance travel.
- Potential resistance from those accustomed to unrestricted personal mobility.

4. **Transformation Scenario: "The Mobility Revolution"**

What if breakthrough technologies fundamentally alter our concept of transportation?

- Widespread adoption of personal flight devices (e.g., jetpacks, flying cars) revolutionizes urban mobility.
- Hyperloop and similar technologies make long-distance ground travel as fast as air travel.
- Virtual and augmented reality advance to the point where physical travel is often unnecessary for work or social interaction.
- Roads become smart, multifunctional surfaces that generate energy, repair themselves, and adapt to changing needs.
- Brain–computer interfaces allow for direct control of vehicles, reshaping the concept of driving.

Implications:

- Radical reshaping of urban landscapes and the built environment.
- Blurring of lines between physical and virtual presence, potentially reducing the need for some types of travel.

CHAPTER 9   SYNTHESIZING FUTURES PRACTICE

- New challenges in regulation and social adaptation to radically new forms of mobility.

These four scenarios present distinctly different visions of how the future of transportation and infrastructure might unfold:

1. The "Continuation" scenario envisions a highly efficient, technology-driven future that's an evolution of current trends.

2. The "Collapse" scenario imagines a future where current systems break down, leading to a more localized, low-tech approach to transportation.

3. The "Discipline" scenario depicts a future of strict regulation aimed at sustainability, significantly changing how and how much we travel.

4. The "Transformation" scenario presents a future where breakthrough technologies fundamentally alter our understanding of mobility and infrastructure.

Each of these scenarios has different implications for society, the environment, and individual lifestyles. They also present different challenges and opportunities for policymakers, urban planners, engineers, and technology developers.

It's important to note that the actual future is likely to contain elements from multiple scenarios rather than perfectly matching any single one. The value of this exercise is in exploring a wide range of possibilities and considering their potential impacts. (See Figure 9-15 as an example).

CHAPTER 9   SYNTHESIZING FUTURES PRACTICE

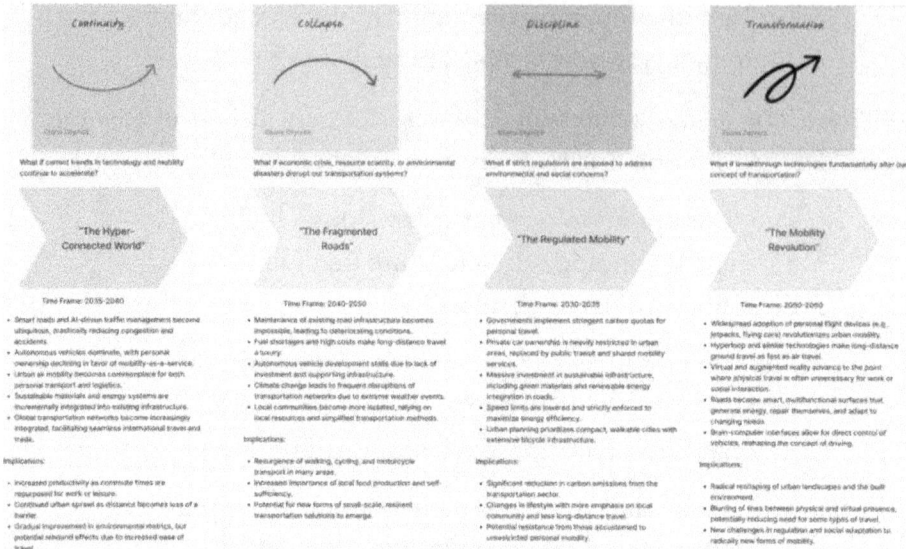

Figure 9-15. *The Four Scenarios for Continuity, Collapse, Discipline, and Transformation*

From here on, it is very important to continue monitoring the environment for potential clues onto what of these archetypes the future is moving toward. With this in mind, now we move on to creating some visual images of the future to better understand its potential implications. It is good to keep in mind that there are many ways to represent the future. Some people prefer to create artifacts from the future, like the work in speculative or critic design, some people prefer to work with narrative and storytelling to project these futures, and some people might decide to create a moodboard, scenario maps, a film, or even 3D works in VR. The ultimate choice depends on the particular circumstances of the study and the needs for communication of results and team participation.

Three options as examples to be used here could be using a 3D engine like Unity or Unreal to create a very customized environment to be projected through a VR or Mixed Reality headset (this option is the most expensive but could yield very impactful results due to the immersiveness

CHAPTER 9 SYNTHESIZING FUTURES PRACTICE

and realistic nature of the experience; a video using real people and real cameras with a storyline to capture the essence of these scenarios (this option might be the second in line in terms of cost); the use of a tool like Runway, which facilitates the creation of video following the story line based on prompt inputs as with the other generative platforms mentioned previously (this option might be more budget conservative); or even the use of Midjourney to generate images that can also aim to provide visuals about each scenario. In this case, Midjourney was used with these preliminary results. (See Figure 9-16 to Figure 9-19).

Figure 9-16. Continuity Scenario

CHAPTER 9   SYNTHESIZING FUTURES PRACTICE

Figure 9-17. *Collapse Scenario*

Figure 9-18. *Discipline Scenario*

CHAPTER 9   SYNTHESIZING FUTURES PRACTICE

Figure 9-19. Transformation Scenario

9.4 Summary

This chapter focused on the synthesis and application of futures methods to develop a comprehensive futures project. Building on previous chapters, it emphasized the importance of grounding futures design within specific cultural, community, and place-based contexts to ensure that the envisioned futures are relevant and resonant with the people and environments they aim to impact. This chapter guided readers through the process of selecting a focal area, such as urban planning, education, or technological innovation, and then applying a combination of futures methodologies—such as scenario building, prototyping, strategic foresight, and backcasting—to create a cohesive and actionable body of futures work.

The chapter highlighted the importance of contextualizing futures work within the lived experiences and values of specific communities, ensuring that the resulting futures are not abstract or imposed but are

245

CHAPTER 9 SYNTHESIZING FUTURES PRACTICE

instead cocreated with and for the people they affect. By integrating methodologies and creating multimedia artifacts, strategic roadmaps, and actionable interventions, readers are encouraged to develop projects that are both imaginative and pragmatic, and at the end of the chapter we presented a tangible case study about the future of the roads and highways infrastructure.

PART IV

AI in Futures Thinking: Opportunities and Challenges

CHAPTER 10

AI and Futures Thinking Methodologies

As we have explored so far, developing resonant visions, scenarios, and artifacts representing potential futures is a crucial practice for cultivating long-term foresight. By rendering speculative possibilities in immersive and experiential ways, we awaken our minds to emerging trajectories and seed new choices about the worlds we'll create through our present actions.

Historically, this vital futures thinking work has relied on human-driven processes—from collaborative worldbuilding workshops to research-intensive scenario modeling. But the rapid advancement of artificial intelligence capabilities is opening up powerful new possibilities for how we imagine, simulate, and choreograph the future.

In this chapter, we will explore the emerging intersections of AI and futures thinking methodologies. We will explore how machine intelligence can be leveraged to augment and scale our abilities to perceive leading indicators, model complex scenarios, and manifest preferable future visions in rich multimedia forms. From environmental scanning and data mining to generative artifact production and simulation modeling, AI is becoming an indispensable cocreative partner in foresight work.

CHAPTER 10   AI AND FUTURES THINKING METHODOLOGIES

At the same time, we will delve into the ethical implications this intersection with AI raises. Speculative futures have historically been bounded by human imagination and cognitive constraints. But as we deploy AI to transcend those limitations in conjuring possible realities, we must be conscious of the new responsibilities and risks involved. We will examine emerging governance frameworks for ensuring AI-assisted futures design does not veer into misuse or hazardous speculation.

10.1 AI in Scenario Development and Analysis

One of the most powerful applications of AI in the field of futures studies is augmenting our capabilities for scenario planning and analysis. As we have previously explored, developing rich scenarios that map out alternative future trajectories is a core practice for cultivating foresight. AI technologies can enhance virtually every phase of this process.

Let's walk through some of the ways AI and machine learning techniques can be synergistically integrated into scenario methodology:

Data Gathering and Environmental Scanning
A first critical scenario input is perceiving the "signals" emerging in the present day that could grow into future disruptions or new realities. This environmental scanning work is turbocharged by AI systems that can continuously ingest, parse, and extract insights from massive, dynamic data flows across scientific journals, patents, social media, news reports, and other unstructured data sources.

Using advanced natural language processing (NLP), computer vision, and other perception AI, these systems can automatically surface nascent patterns, anomalies, and forces percolating that could seed wider impacts down the line. The AI handles the data firehose so human analysts can focus on higher-order signal prioritization and futures modeling from a more comprehensive informational landscape.

Scenarios Signal Processing

Once key emerging signals are identified through AI-powered scanning, that data can be fed into machine learning models to further enrich and connect the dots into scenario inputs. For example, large language models could be fine-tuned to automatically generate reports analyzing the intersecting impacts of perceived technological and cultural signals on future economic, political, and social systems.

AI could also be used to surface less obvious connections between signals by applying unsupervised learning to mine for higher-order thematic patterns spanning the multidisciplinary data. These enriched overviews feed into the subsequent scenario development phases.

Dynamic Scenario Modeling and Simulation

Developing coherent, interrelated scenario narratives has traditionally been a laborious process of joining expert beliefs with foundational data. But new advances in AI like large language models, knowledge graphing, and multiagent simulation are allowing scenario developers to generate and dynamically model future scenarios at unprecedented scales.

Using frameworks like Constitutional AI, these systems can autonomously assemble initial scenario outlines and renderings by ingesting the environmental data, then iteratively expanding and recombining the elements through cycles of open-ended generation. Developers can then edit, ideate around, or regenerate the AI scenario sketches, creating a feedback loop of human–AI cocreation.

In multiagent simulations, the AI models different influencing forces like technologies, policies, behaviors, etc., as autonomous agents with their own objectives and rules. By simulating their interactions within different contexts modeled on the input signals, the AI essentially authors dynamic scenario storylines of how events could unfold and impacts cascade based on those starting conditions.

CHAPTER 10   AI AND FUTURES THINKING METHODOLOGIES

Enriching Scenarios with Immersive Artifacts

Beyond the narratives, AI technologies can create multimedia artifact renderings and simulations to make scenarios experientially immersive. Using text-to-image, code-to-video, and other generative AI, developers can depict the qualitative textures and artifacts of future scenario worlds.

From speculative products and interfaces to simulated environments and cultural experiences, the automated artifact generation allows immersion into the felt qualities of how a given scenario's future could manifest. This multimedia augmentation enriches the sensory presence of scenarios, making them powerfully resonant beyond just dry analytical descriptions.

Note that automated generators are supplements, not substitutes, for embodied prototyping. While they accelerate exploration, they can also flatten materiality, over-polish visuals into a 'model style,' or echo dataset biases. When sensory richness matters, prefer a hybrid workflow: use AI for quick ideation and variation, then translate promising directions into crafted physical or high-fidelity prototypes, and validate with audience and sensory criteria (sound, texture, scale, ergonomics).

Scenarios Analysis and Insight Extraction

Finally, AI can be leveraged to analyze scenarios—processing the narrative data and simulation outputs to surface insights into implications, dependencies, and strategic tensions. Using machine learning models trained on relevant domain datasets, the AI could automatically highlight key uncertainties, leverage points, or potential issues contained across different scenario versions.

Essentially, the AI becomes a cognitive support system for human foresight teams, enhancing different scenario process phases with automated information processing, synthesis, and pattern detection capabilities that would otherwise be extremely labor-intensive and time-consuming, if not impossible, for limited human analysts.

Of course, there are some key watch outs when integrating AI into high-stakes futures processes. We must be vigilant against AI bias, hallucination, and artificial scenario echo chambers forming from opaque

CHAPTER 10   AI AND FUTURES THINKING METHODOLOGIES

feedback loops in the generative models. Rigorous human oversight and multidisciplinary input is crucial for ensuring AI-generated scenarios are coherent, plausible, and map meaningfully onto reality.

But when applied judiciously, AI stands to become an indispensable liberating force multiplier for futures thinking and scenario development. Rather than spending countless human labor hours chasing futures signals and manually iterating scenarios, we can work in symbiosis with AI to scale our collective foresight capacities. The goal is leveraging AI to transcend bottlenecks currently constraining both the speed and aperture of our civilization's futures vision and preparedness.

10.2 Case Studies: AI-Driven Futures Thinking and Speculative Design

There is this company I recently learned about called Shaping Tomorrow that is pioneering the use of AI for automated foresight work. Their core product is an AI system specifically designed to supercharge the environmental scanning and horizon scanning process.

Instead of teams of analysts spending countless hours manually sifting through research publications, patents, news reports, and other data sources looking for signals of change, Shaping Tomorrow's AI does the heavy lifting. It continuously ingests massive datasets across pretty much every domain you can think of—scientific research, tech trends, political developments, socio-cultural shifts, you name it.

Then using natural language processing and other AI techniques, it automatically surfaces patterns and clusters of insights that could indicate emerging issues or discontinuities on the horizon. Essentially, it's like having an ultra-powered cognitive prosthetic for perceiving very faint ripples of change from the peripheral edges before they grow into tidal waves.

253

CHAPTER 10   AI AND FUTURES THINKING METHODOLOGIES

What is interesting is how the AI can connect disparate data points from wildly different contexts into coherent narratives around potential futures scenarios. I was blown away when their system identified a biotech patent coupled with some fringe community behaviors documented by anthropologists as signals converging into a plausible future pathway around democratized bioengineering. The kind of insight that would likely have been missed through traditional human-driven scanning methods.

Of course, the AI's output is still just a first-staged analysis that then needs to be pressure tested and enriched by human foresight professionals. But it gives such a cognitive headstart by surfacing those initial threads of emergent change ready to be unpacked and woven into deeper narratives and scenarios.

The team at Shaping Tomorrow sees their AI as almost like an automated curator—one that can continuously ingest and parse vast terrain maps of incoming signals, while flagging areas of interest for the human foresight teams to then explore with rich cultural context, strategic insight, and participatory artifacts. A way to scale the often tedious data filtering work, freeing up more cognitive bandwidth for the interpretive and imaginal work.

From what I have seen, Shaping Tomorrow is really pushing the boundaries of how AI can start automating the "view" aspect of foresight—that impartial triaging of information terrains to surface threads and clusters of emergent change ready for human-guided sensemaking and scenario development. It's an AI system specifically optimized as an intelligent sensor for perceiving the faint signals of futures emerging all around us.

I could see capabilities like this being integrated into all kinds of strategic planning and innovation pipelines—having AI sentries on overwatch for disruptions, then cueing up the human foresight teams to unpack those signals into multimedia scenario work. The symbiosis of automated mass data sifting coupled with human contextual mastery and futures design feels like it could really move our foresight capacities into a whole other gear.

That said, there are still valid concerns I have around the boundaries and failure modes we need to be vigilant about with this kind of AI augmentation for foresight work. Particularly around the risks of codifying human biases, blind spots, and narrow framing into the AI's prioritization and signal clustering models.

If the training data and foundational assumptions we bake into the AI's aperture are too constrained by industrialized worldviews, capitalist realism, or human separatist conceits, then we could very well end up with an "intelligent sensor" that is inherently terrible at actually perceiving the deepest, most pivotal signals of transformative futures emerging.

There's a real danger that the AI's information terrain triaging and "area of interest" flagging for human teams ends up reproducing existing institutionalized biases and optimizing for incremental, human circumscribed readings of change—rather than surfacing the soft casualties of radically contingent alternative trajectories.

We could end up with an "automated curator" that remains deafly blind to the subtle wisdoms, pluralistic narratives, and imaginal dynamics seeding profound metamodern transformations. One that stubbornly privileges the known and quantifiable while missing the obscured openings.

This could lead to a pernicious imbalance, where the AI is clumsily over-indexing our foresight work on loud, obvious permutations of industrial/capitalist novelty like new product cycles and tech gizmos. While missing the textured, indigenous wisdoms and re-integrative undercurrents arising from the margins as generative sources of societal revisioning.

So while I'm very excited about the potential of AI to radically augment our environmental scanning, signal processing, and data synthesis capabilities for foresight work, we must be extraordinarily prudent about bounding that augmentation with pluralistic, metamodern counterweights.

The future-scanning AI can't just be a brute force quantifier churning through datasets in service of technocratic apertures. It must have embedded, continuous integration points for subjective/subversive imaginaries, decolonial/Earth-centered re-integration narratives, and heterogeneous futuring practices to keep flexibly attuning its gaze.

We should be disciplined about complementing the AI capabilities with living frameworks for bottom-up, participatory attunement and kaleidoscopic reframing. Community story work, applied metaphor surfacing, pluriversal priming—all of these "metamodern judo" moves become obligatory practices when working with AI to truly expand our collective futures perception rather than narrow its peripheries.

Cultivating this symbiotic counterbalance, with dynamic human-community counterweights continually reintegrating and reinflecting the AI's apertures, is how we can truly elevate our futures literacies rather than ossify in the sedimentary biases of our industrial-capitalist tendencies.

An AI pipeline for signal processing and information terraining could be immensely powerful—but only if it remains exquisitely beholden to the interpretive, imaginative, and integrative arts that awaken our pluralistic foresight rather than hardening its blinders.

10.3 Ethical Implications of AI in Design

One of the core issues that is common while using AI systems is the AI "hallucination" or generation problem—the potential for AI models to output incoherent, nonsensical, or harmful artifacts when prompted to imagine new concepts outside their training distributions. Just as current large language models can produce both profound and sometimes even nonsense or made up text when pushed far enough, generative AI crafting immersive scenarios or speculative designs could proliferate distorted, faulty, or ungrounded visions.

There are already examples of text-to-image models producing racist, violently skewed, or deeply unsettling imagery when prompted in ambiguous ways. As we scale up the AI systems constructing entire future world renderings, the stakes of such distortions and aberrations amplify exponentially. Unintended artifacts birthing skewed metaphysical premises or nightmarish ontological slates could become an existential risk.

Then there are the bias issues—the tendency for AI systems to replicate and even amplify the demographic skews and contextual blind spots inherent to their training data. Using current AI to speculatively model futures runs the risk of simply regurgitating the status quo filtered through the overwhelmingly Western/Educated/Industrialized/Rich/Democratic (WEIRD) perspectives dominating datasets and internet corpora.

We could end up with AI proliferating speculative scenarios constrained by anthropocentric, colonial, and extractive premises—visions that fail to transcend the root paradigms driving our planetary polycrisis. The uncanny valley expands to an existential bias-field propagating the very frames in need of reinvention.

There are also accountability vacuums—with AI models producing artifacts autonomously at scale based on fuzzy prompts. For reflection, who takes responsibility for harmful premises, concepts, or worldviews proliferated? The developers? End-users? The models themselves if they become sovereignly self-configuring?

These fault and accountability risks are not theoretical. We are already seeing them emerge in AI deployment today, and that is just with systems generating language or static images. As we compose richer, experientially immersive models of whole future ontologies, issues of harm, bias, and ungrounded worlding could scale nonlinearly.

That said, there is also a profound opportunity here. If we can find ways to responsibly train and constrain AI futures design systems with the proper ethical guard-rails, multivocal perspectives and "human-in-the-loop" accountability frameworks, they could become catalytic engines for transcending our paradigmatic biases at scale.

We could seed models with intrinsically antiracist, decolonial, planetary-centric premises. This would activate its generative capacities from profoundly new, liberating vantage points that circumvent Western/human supremacy distortions.

From there, with robust oversight and accountability frameworks, the AI's infinitely generative capacity could become a kind of planetary metamodeling engine, an ability to prototype possible integrated futures at scales and velocities far exceeding the abilities of any siloed human team. We could then perform the reality check and elevate the resulting prototypes for further refinement.

As it is evident, these are enormously complex challenges humanity is just beginning to grapple with. Figuring out how to create AI systems capable of ethical, grounded worlding and futures prototyping at scale, may be one of the great threads of creative and existential work ahead.

Pioneers in this arena will need to blend cutting edge AI capabilities with deep wells of cross-disciplinary wisdom, looping in indigenous futurists, cosmic philosophers, complexity thinkers, evolutionary systems architects, and more. The goal is a kind of bounded pluralism: systems that actively seek multifocal perspectives, including marginalized, Indigenous, Global South, disability justice, and environmental stewardship proxies, while explicitly excluding violent, rights-denying, or supremacist ideologies. Perspectives are weighted transparently using human-rights norms, distributive and procedural justice, and planetary boundary science, so the model widens its lens without amplifying harm.

10.4 Summary

This chapter explored the powerful intersection between artificial intelligence (AI) and futures thinking methodologies, highlighting how AI can enhance scenario development, environmental scanning, and speculative design. By leveraging machine learning and AI-driven tools,

CHAPTER 10   AI AND FUTURES THINKING METHODOLOGIES

futures practitioners can process vast data streams, identify emerging trends, and dynamically model complex scenarios more efficiently. AI's ability to generate immersive multimedia artifacts, simulate future environments, and analyze scenario implications enables a more expansive and detailed exploration of potential futures. However, human oversight remains essential to ensure the coherence, relevance, and ethical grounding of AI-assisted futures work.

The chapter also delves into the ethical implications of using AI in futures thinking, emphasizing risks such as AI-generated bias, "hallucination," and the perpetuation of harmful societal paradigms. While AI offers immense potential for transcending human cognitive limits, it can also proliferate skewed or damaging speculative scenarios if not carefully managed. To harness AI responsibly in futures design, the chapter advocates for robust ethical guardrails, accountability frameworks, and the inclusion of diverse perspectives in AI training and development. This approach ensures that AI is used not only to expand foresight capabilities but also to create equitable and inclusive futures.

PART V

Advanced Concepts and Methodologies

CHAPTER 11

Interactive and Transdisciplinary Approaches with AI

As AI capabilities exponentially advance, they are catalyzing profound shifts in how we conceptualize, prototype, and ultimately manifest our visions into reality. AI is becoming an indispensable collaborative partner across the entire cycle of creativity, analysis, and construction.

In the field of interactive media design, disciplines like UI/UX, gaming, digital storytelling, and more are being radically reshaped by AI's cognitive augmentation. Machine learning models can now engage in open-ended ideation, rapidly prototyping novel interfaces, narratives, and virtual worlds through iterative cocreation with human designers.

AI is also enabling richer interactivity, personalization, and context-aware experiences that can dynamically adapt and evolve based on the user's emotional or psychological states, environment, and needs. Intelligent digital experiences are transcending static, "one size fits all" formats.

But the impacts of AI extend far beyond digital realms into catalyzing transdisciplinary design integration with other sectors. We are witnessing the blurring of boundaries between design and fields like computer science, engineering, biotechnology, urban planning, materials, and science to name a few.

CHAPTER 11 INTERACTIVE AND TRANSDISCIPLINARY APPROACHES WITH AI

AI is becoming the connective fabric allowing seamless integration of ethical, experiential, and aesthetic human-centered principles into fields previously dominated by raw technological capabilities. Design thinking is shaping the very constructs and embedded values of transformative technologies as they are being developed.

For example, AI models can now interface directly with engineering simulations and CAD/CAM environments, allowing designers to quite literally "sketch" physical products and architectural forms with machine learning acting as an intelligent cooperative partner translating human intentions into viable constructions.

This transcends crude technological determinism, embedding human ingenuity and creativity at the core of innovation pipelines rather than seeing human designers as downstream receivers of new technological capabilities decided elsewhere.

Similarly, AI's prowess in domains like materials science is unlocking new frontiers for product and industrial design. We can now sculpt with AI-optimized metamaterials and living matter, embedding dynamic properties, programmable behaviors, and biointelligence into physical artifacts that were previously inert.

From biodigital buildings that can adapt, self-repair, and harmonize with their ecosystems to apparel and consumer products with dynamically transforming forms and functionalities, designers are harnessing AI to transcend static objects for designing with living, evolving systemic assemblages.

CHAPTER 11   INTERACTIVE AND TRANSDISCIPLINARY APPROACHES WITH AI

11.1 AI in Interactive and Digital Media Design

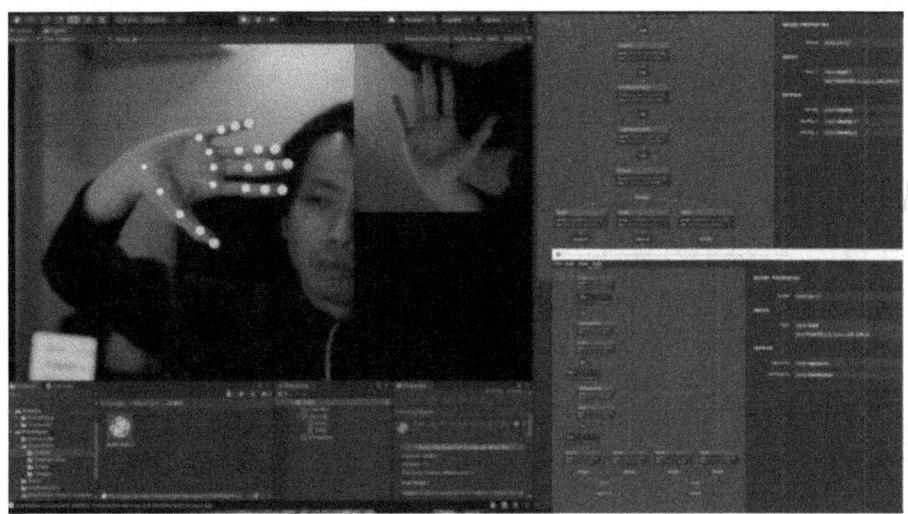

Figure 11-1. *Flexible AI Model Implementation Unity Sentis*

One of the most visible and transformative frontiers where AI is augmenting human creativity is the field of interactive digital media design—encompassing disciplines like user experience (UX), user interface (UI), video game development, digital storytelling, and immersive virtual/augmented reality. (See example on Figure 11-1).

At a fundamental level, AI is supercharging interactive designers' abilities to ideate, prototype, and develop novel digital experiences at an unprecedented speed and complexity. Machine learning models can now engage in open-ended cocreation loops, allowing designers to offload tedious tasks while amplifying their creative bandwidth.

CHAPTER 11   INTERACTIVE AND TRANSDISCIPLINARY APPROACHES WITH AI

For example, AI storytelling models can now improvise branching interactive narrative arcs, dynamic environments, and behavioral profiles for non-player characters (NPCs) based on high-level human prompts. Rather than designers hand-scripting every narrative permutation, the AI becomes a cognitive multiplier—rapidly filling out possibility spaces while grounding the outputs in human-specified creative constraints.

This allows game developers and digital storytellers to push the boundaries of what is possible in terms of open-ended, interactive worldbuilding within their creative visions. The AI handles the onerous task of pondering every potential thread while designers focus on crafting the higher level narrative anchors and tonal qualities.

Similarly in UI/UX design, generative AI models as illustrated in Figure 11-2 below, can engage in back and forth ideation around interface concepts, screen flows, microinteractions and more, rapidly prototyping multiple design variants for human designers to critique or combine into optimal solutions. Essentially, the AI proposes uninhibited ideas that the human designer then masks through an iterative feedback loop.

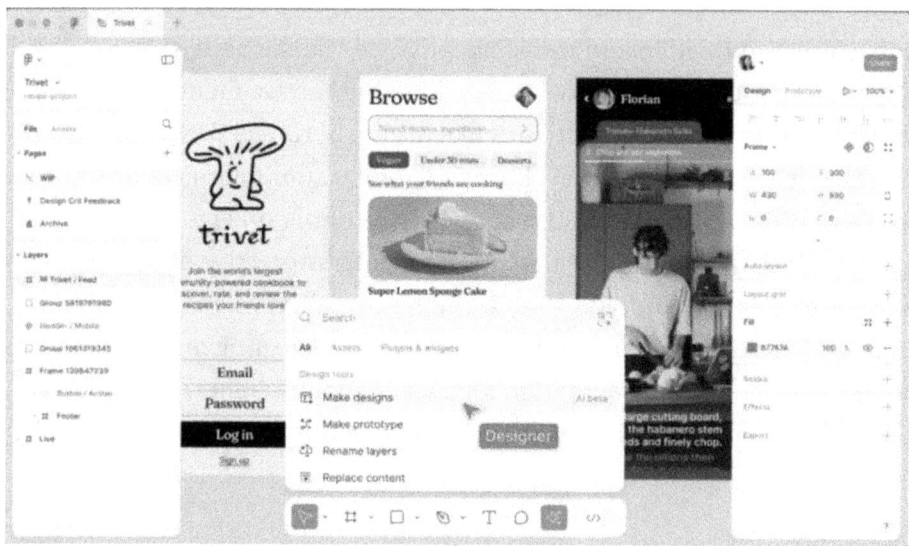

Figure 11-2. *Figma AI*

CHAPTER 11 INTERACTIVE AND TRANSDISCIPLINARY APPROACHES WITH AI

Generative AI is also being leveraged to create dynamic, data-driven visualizations and information architectures that can intelligently adapt their styling, content ordering, and even the interactions themselves based on the user's context, inputs, and behavioral profiles. This transcends static, one-size-fits-all UI/UX approaches.

For example, an AI-powered media app could customize its layout, visualizations, and available actions not just for different user profiles but for the same user's shifting mindsets and needs in different situational or emotional contexts over time.

In gaming, AI is enabling not just richer non-player character behaviors, but entirely AI-generated interactive environments that can dynamically evolve in complexity and narrative cohesion as the human player makes choices. The AI world-simulation responds intelligently to player actions while continuously maintaining interactive authenticity and story logic. (See Figure 11-3).

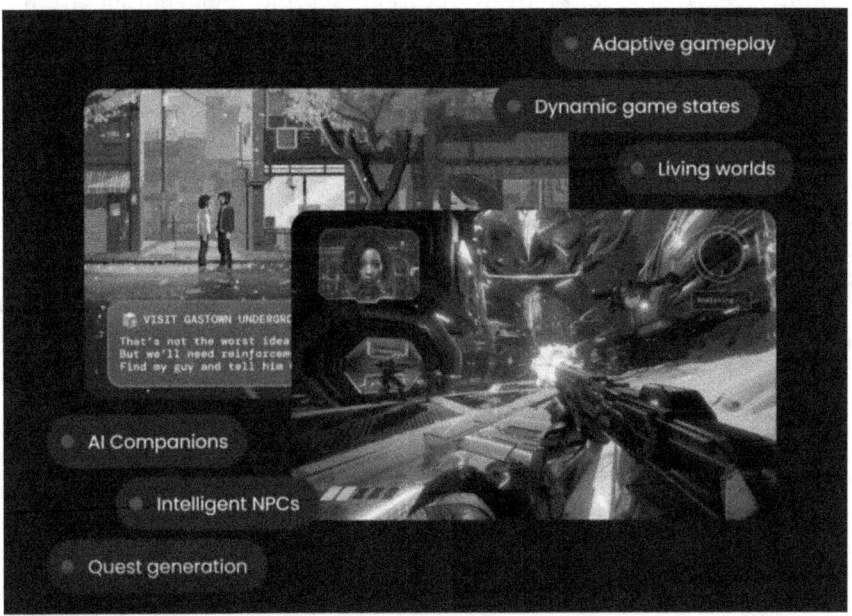

Figure 11-3. Inworld AI

CHAPTER 11   INTERACTIVE AND TRANSDISCIPLINARY APPROACHES WITH AI

Similarly, for virtual and augmented reality developers, AI is becoming essential for crafting immersive simulated worlds and experiences that can engage human perception and cognition in fundamentally adaptive ways.

Rather than VR/AR environments being statically rendered, AI models can imbue them with responsiveness—changing the lighting, physics, spatial mappings, and narrative elements based on tracking the user's gaze, emotional states, decisions, and even physiological data like heart rate in real-time. The virtual world intuitively shapes itself around the human participant.

Embedding AI into interactive digital media also raises ethical considerations around value alignment, transparency, and user privacy/security. For reflection, if an AI system is shaping interactive experiences and environments in unconstrained ways, how do we ensure it does not reinforce harmful biases, distribute manipulative narratives, or illegally surveil end users?

There is a growing need for human-centered AI governance and multi-stakeholder collaboration to establish clear ethical guardrails as these intelligent cocreative systems become further enmeshed into experience design pipelines and user-facing applications.

But when leveraged with rigorous principles, AI allows interactive designers to transcend the production constraints and creative bottlenecks of industrialized content. We can craft rich, adaptive digital worlds that engage users as coparticipants and coauthors intuitively guiding their own journeys—enabled by intelligent systems with infinite generative bandwidth.

11.2 AI's Role in Blending Design with Other Disciplines

One of the most transformative impacts of AI is its ability to bridge traditional disciplinary silos, catalyzing transdisciplinary integration between design and sectors like engineering, computing, materials

CHAPTER 11   INTERACTIVE AND TRANSDISCIPLINARY APPROACHES WITH AI

science, biotechnology, urban planning, and more. AI is becoming a connective fabric that allows human-centered design principles to be seamlessly encoded into other technical disciplines.

In fields like engineering and computer science that have historically been driven more by raw technological capabilities, AI is opening up new avenues for design thinking to become a core part of innovation processes from the start. Rather than designers receiving finalized technological systems to apply UX principles after-the-fact, AI allows them to essentially "co-design" the fundamental architectures and embedded properties.

For example, AI models can now directly interface with engineering simulation and CAD environments. Designers can sculpt physical product forms and architectural constructions using intuitive tools, with AI translating their sketches and real-time inputs into viable technical specs, material constraints, production requirements, and more. It allows designers to rapidly ideate and iterate on physical artifacts through a tightened feedback loop with simulation rather than disconnected handoffs.

Similarly, with AI catalyzing tighter integration between design and emerging domains like bioengineering and materials science, designers leveraging AI can move beyond traditional manufacturing with inert materials and instead sculpt with metamaterials, biosynthetics, and even living matter imbued with intelligent properties.

Initiatives like Koniku Kore use AI with neuroscience, neurotechnology, and biotechnology to create "smell cyborgs," devices that combine live biological cells with silicon to detect and analyze volatile organic compounds in the air. Their technology uses engineered proteins in neurons to sense and process smells, aiming to diagnose diseases and maintain health in real-time.

Researchers have created AI systems that can optimize microstructures, lattices, and molecular compositions of materials based on desired behaviors, enabling product designers to craft self-assembling,

CHAPTER 11   INTERACTIVE AND TRANSDISCIPLINARY APPROACHES WITH AI

dynamically transforming, and biodegradable artifacts that were previously impossible. This opens up entirely new design spaces of biocompatible, evolving, and self-repairing product ecosystems.

In architecture and urban design, AI is empowering transdisciplinary convergence with fields like ecology, climate science, and sociology. Architects can now leverage AI to co-optimize their building and city designs in real-time based on complex environmental modeling, citizen well-being analytics, generative zoning rules, and more.

Across sectors, a common shift is AI augmenting and densely integrating design across disciplines—and beyond a solely human-centered lens. Designers are working with communities, ecologies, and technical systems as coparticipants, becoming coauthors of innovation trajectories with AI as a collaborator and more-than-human contexts as stakeholders.

This includes bio-hybrid practices such as *cyborg botany, which involves pairing plants with sensors/actuators to sense and respond to microclimates, alongside multispecies design that considers habitats, nonhuman life, and infrastructures as design assets. These approaches sit alongside service, product, and policy work, expanding who and what design is "for"cross sectors; the common theme is AI augmenting and densely integrating design thinking across disciplines rather than leaving it quarantined in a silo. Designers become coauthors and creative directors shaping innovation trajectories with AI as a collaborator.

Cyborg botany: Bio-hybrid systems where plants interface with electronics to sense/act (e.g., urban heat sensing, responsive irrigation, or air-quality signaling).

Importantly, this new era of AI-augmented transdisciplinary design requires evolved mindsets and skillsets from designers themselves. They must become adept in new AI tooling, computational techniques, and specialized knowledge domains (e.g., materials science, bioinformatics, etc.)—blending these with their classical design foundations.

11.3 Interviews with AI and Design Industry Leaders

Interview of John Maeda by Diana Olynick

Interviewer: In your work, you often explore the intersection of technology, design, and human experience. How do you see the principles of simplicity playing a role in this intersection, particularly in creating sustainable and human-centric design with the advent of artificial intelligence?

John Maeda: The principles of simplicity are crucial when balancing complexity and accessibility. Simplicity allows more people to adopt new technologies and designs, reducing barriers to entry. Complexity, while sometimes necessary and interesting, can be a hindrance. Over time, I have learned that the more experienced you become, the better you get at simplifying complex concepts. An example from my own experience is a famous design professor, Wolfgang Weingart, who would simplify his lectures on typography every year, making them easier to understand while maintaining their depth. This journey from complexity to simplicity is essential in making technology more human-centric and accessible.

Interviewer: You have mentioned that art is about asking unanswerable questions. How do you think this approach can contribute to a more ethical and considerate use of AI in design?

John Maeda: Art and the humanities promote critical thinking, allowing for deep reflections on the implications of creation without the pressure of immediate application. This critical perspective is vital for ethical considerations, as it provides the space to question the impacts of AI beyond its functional capabilities. In contrast, business and industrial applications focus predominantly on making and implementing solutions, often leaving little room for ethical questioning. By integrating the artistic approach of asking unanswerable questions, we can ensure a more thoughtful and ethically sound use of AI in design.

CHAPTER 11 INTERACTIVE AND TRANSDISCIPLINARY APPROACHES WITH AI

Interviewer: Your career spans various roles from academia to corporate leadership. How has this journey influenced your perspective on the role of technology in enhancing human well-being?

John Maeda: My diverse career has underscored the importance of understanding and communicating the essence of technology, particularly computer science, which is often invisible to the layperson. In my book *How to Speak Machine*, I spent years demystifying computer science, making it accessible and understandable. This journey taught me that technology has immense potential to automate both beneficial and harmful systems. The challenge lies in balancing these forces to move society forward thoughtfully. Being able to see both the sides of technology—its potential and its pitfalls—is crucial for leveraging it to enhance human well-being.

Interviewer: As AI continues to evolve, what ethical considerations should designers keep in mind to ensure that technology serves to augment human creativity rather than replace it?

John Maeda: One major consideration is accountability. Machines and AI systems cannot be held responsible for their actions; the responsibility lies with the humans who design and implement these systems. As AI becomes more sophisticated and human-like, there is a tendency to anthropomorphize it and deflect blame onto the technology itself. This is a dangerous mindset. Designers must ensure that the systems they create are transparent and that the responsibility for their actions remains human. This accountability is essential to using technology to augment rather than replace human creativity.

Interviewer: What excites and concerns you most about the current advancements in AI?

John Maeda: I am excited about AI's potential to revolutionize various fields, but I am also concerned about the general lack of understanding surrounding it. Even for technical experts, grasping the full scope of modern AI is challenging. This disconnect can lead to misinformed decisions and unrealistic expectations. Therefore, it's crucial for more people to educate themselves deeply about AI to harness its potential responsibly.

Interview of Alice Rawsthorn by Diana Olynick

Interviewer: The Design Emergency project investigates design's role in creating positive change. Can you discuss a specific case from this project that exemplifies the transformative power of design in tackling contemporary issues?

Alice Rawsthorn: Design Emergency has created an incredible opportunity for Paola Antonelli, senior curator of architecture and design at MoMA, New York, with whom I co-founded the project, and I to investigate and share the work of designers, architects, engineers, activists, farmers, technologists, and others, who are pioneering new ways of addressing the complex, often terrifying challenges we face at this perilous, intensely volatile time. Immersing ourselves in the work of these remarkable designers has been thrilling and empowering.

All the projects we have explored through Design Emergency have been extraordinary in their own ways, but one of my favorites is a truly epic design endeavor, the Great Green Wall of Africa. This is a long-term, pan-African project that began in 2007 and is run collectively by 21 countries in the Sahel region with the aim of repairing the damage caused by climate emergency to one of the hottest, driest, poorest parts of the world by cultivating an 8,000 km, or 5,000 mile, strip of vegetation across the southern edge of the Sahara from Senegal on Africa's west coast to Djibouti in the east.

The Great Green Wall is an inspiring example of the use of strategic design to tackle a massive, intersectional problem whereby decades of drought have caused crop failure, famine, soil erosion, and desertification, which have then fueled socio-economic problems of poverty, conflict, and mass migration. Like many of the new genre of epic design endeavors, it is long term, complex, and often controversial.

CHAPTER 11   INTERACTIVE AND TRANSDISCIPLINARY APPROACHES WITH AI

Each country is free to regenerate its land as it wishes. Senegal has focused on tree planting, and Burkina Faso on reviving ancient planting and irrigation techniques. Progress has been erratic, not least as individual countries have been beset by economic, political, and health crises. Both Ethiopia and Niger were doing well until civil war erupted. Three years ago, less than 20% of the wall was completed. But a group of donors led by France and the World Bank then pledged $19.5 billion to the project, giving the Great Green Wall a fighting chance of completion and of achieving its objective of creating ten million new green jobs in the Sahel region by 2030.

Not only would this have a profound impact on the lives of millions of people living in that region, it could transform perceptions of design by demonstrating its power as a strategic tool to tackle major challenges. Completing the Great Green Wall could be a game changer for design, as well as for this deeply vulnerable part of the African continent.

Interviewer: In your book *Design Emergency: Building a Better Future*, you explore how design can address complex social, political, and ecological challenges. How do you see the role of design evolving in the age of AI, and what unique opportunities does AI present for designers?

Alice Rawsthorn: Design is a complex and elusive phenomenon that has adopted many different meanings over the centuries. But it has always had one elemental role as an agent of change, which can help to ensure that changes of any type—social, political, economic, scientific, cultural, whatever—are interpreted in ways that will affect us positively, not negatively. And one of design's core functions throughout its long, complex history has been to analyze and anticipate the probable impact of new technologies to enable us to make the most of their benefits, while avoiding their dangers.

This applies to AI, just as it has to all the once-new technologies that preceded it. We all know that advances in AI have the potential to produce significant benefits. Improving agricultural yields, for example, by enabling farmers to more accurately analyze the quality of their land, the weather patterns that affect it, and to reduce their use of pesticides without impeding efficiency. Or helping public health care services to operate more effectively by using large language AI models to identify smarter ways of deploying specialist staff and allocating resources, such as post-surgical beds, while analyzing scans and x-rays with greater accuracy.

These benefits, and many more, could have a positive impact on the lives of hundreds of millions of people in both developed and developing economies, if—and it's crucial if—the application of AI is designed with due care, thoughtfulness, and precision. If it isn't, such a potent technology could have devastating consequences, by accelerating scientific fraud, for example, or by intensifying the data bias that has already caused serious problems by aggravating the impact of misogyny, racism, ableism, ageism, and other prejudices in AI-driven surveillance systems and job applications.

Such risks make it even more than usually important that AI systems are designed wisely and sensitively. Equally important is for them to be designed from an anticipatory perspective so that potential problems can be identified and quashed or, at the very least, minimized from the outset. The growing significance of anticipatory design is a compelling new challenge for designers to address.

Interviewer: Your TED talks and participation in global events have reached a wide audience. How important is public engagement in your work?

Alice Rawsthorn: Confusing and cliché-ridden though it often is, design is a ubiquitous force that affects every aspect of all our lives. It is impossible for us to ignore it, but we can try to ensure that our design decisions are as intelligent and enlightened as possible, and to do that we need to try to understand design and its impact on us. For me, this is a powerful incentive to try to ensure that my work reaches as wide an audience as possible.

Another factor is to try to ensure that the politicians, investors, NGOs, civil servants, entrepreneurs, and others who have the power to greenlight, or scrap, powerful and ambitious design projects are as well informed as possible about design. How else can they be expected to make the most of its benefits, and to avoid its dangers? Unless this happens, design will never achieve the public confidence and political support it needs to realize its full potential as a force for positive change at this deeply turbulent time.

Alice Rawsthorn is an award-winning design critic, the author of books including Design as an Attitude and Hello World: Where Design Meets Life, and cofounder of Design Emergency.

Interview of Jerome Glenn by Diana Olynick

Interviewer: Given your extensive experience with futures research and global trends, how can organizations better manage uncertainty and volatility when creating long-term strategies and designing for future scenarios?

Jerome Glenn: To manage uncertainty and volatility effectively, organizations need to think beyond trying to predict the future. It's about creating systems that adapt and respond as new information emerges. I've often stressed the importance of real-time foresight, which allows organizations to continually gather insights and adjust their strategies. This isn't just a one-off exercise; it's a constant feedback loop. You have to build flexibility into your approach from the beginning. That's why methods like the Real-Time Delphi are crucial—they enable you to stay responsive to change.

Moreover, organizations should develop multiple scenarios to account for a range of possible futures. It's not just about preparing for the most likely future but thinking about the outliers—the unexpected events that could dramatically shift the landscape. The Futures Wheel that I developed helps map out those ripple effects, enabling you to see how one change can lead to cascading impacts. This helps you prepare for broader consequences, not just the immediate ones.

CHAPTER 11   INTERACTIVE AND TRANSDISCIPLINARY APPROACHES WITH AI

A key part of this is also collaboration. No one organization can fully understand global challenges in isolation. By bringing in diverse perspectives—whether from governments, businesses, or civil society—organizations can gain a more nuanced understanding of global trends. This is something we consistently emphasize at the Millennium Project. The ability to tap into different viewpoints allows for more comprehensive foresight and better decision-making.

Lastly, emerging technologies like AI can assist in managing this complexity. AI can process vast amounts of data, helping us recognize patterns that might not be immediately obvious to human analysts. As we move closer to AGI, these technologies will not only help forecast but also potentially recommend actions based on those forecasts. It's essential to leverage these tools as part of an integrated foresight system.

In short, the goal isn't to control volatility but to be adaptive and resilient in the face of it. That's where strategic foresight becomes a real asset—it's not about predicting the future; it's about being prepared for whatever comes.

Interviewer: What are the potential impacts of artificial general intelligence (AGI) on strategic foresight and futures design, particularly in shaping long-term scenarios and planning?

Jerome Glenn: The introduction of Artificial General Intelligence (AGI) could fundamentally transform strategic foresight and futures design. AGI, unlike current Artificial Narrow Intelligence (ANI) systems, will have the ability to autonomously learn, reason, and solve novel problems across a wide range of domains—essentially doing what humans do, but potentially better and faster. This level of intelligence will have significant implications for how we approach long-term planning and scenario building.

First, AGI could drastically improve the precision and depth of scenario analysis. Current foresight methods rely heavily on human experts and narrow AI to identify trends and weak signals. With AGI, the ability to analyze vast amounts of complex data from multiple domains simultaneously would be exponentially enhanced. AGI could identify correlations and insights across global systems—politics, economics, technology, environment—that are difficult for humans to detect. It could rapidly generate and evaluate a much broader range of scenarios, helping organizations anticipate more precise long-term outcomes and potential risks.

From a governance perspective, one of the concerns, as noted in the Millennium Project's AGI Governance Reports (Phases 1 and 2), is that AGI will change the nature of decision-making itself. AGI will be able to offer dynamic scenario updates in real time, adapting as new information comes in. This will enable decision-makers to adjust strategies on the fly, rather than relying on static, periodically updated scenarios. Such responsiveness could be crucial in volatile, fast-changing global environments.

However, the introduction of AGI also poses risks. Unregulated AGI could exacerbate existing geopolitical tensions and lead to the emergence of unpredictable, competing strategies. If multiple AGIs are developed by different countries or corporations without global coordination, their interactions could create scenarios that are beyond human comprehension or control, leading to what some experts call "emergent behaviors" that may be detrimental to human goals. This highlights the need for global governance mechanisms to ensure that AGIs align with shared human values and goals, as discussed extensively in our reports.

Interviewer: Based on your work with the "State of the Future" reports, what are the most critical emerging trends and challenges that organizations should be aware of when designing for the future?

Jerome Glenn: When we look at the "State of the Future" reports, several critical trends stand out that organizations should consider when designing long-term strategies. One of the most important is technological acceleration. We're seeing exponential growth in fields like AI, biotechnology, and nanotechnology, all of which have the potential

CHAPTER 11   INTERACTIVE AND TRANSDISCIPLINARY APPROACHES WITH AI

to drastically change industries. For example, AI and robotics are already reshaping labor markets, potentially leading to both significant unemployment and new job creation in different sectors.

Another pressing issue is climate change. The increasing severity of climate-related disasters, from rising sea levels to more frequent extreme weather events, demands that organizations incorporate sustainability into their core strategies. The cost of inaction could be catastrophic, not just for the environment, but for businesses themselves.

Demographic shifts also play a significant role. We're facing aging populations in some parts of the world, while others experience youth bulges, particularly in developing nations. This creates a complex dynamic for the future workforce, with both skilled labor shortages and surpluses affecting different regions.

Moreover, geopolitical instability is growing. The balance of global power is shifting, with countries like China and India rising, which brings both challenges and opportunities for international businesses. Organizations need to be prepared for a more multipolar world, where political and economic power is more widely distributed.

Finally, inequality—both economic and digital—remains a persistent challenge. As technological advances continue, those who lack access to these innovations risk being left further behind, exacerbating social unrest. Organizations have a responsibility to ensure inclusive growth and equitable access to technology and resources.

Interviewer: How can governments and international organizations effectively integrate futures research into their policy-making processes to address long-term global challenges?

Jerome Glenn: Governments and international organizations can integrate futures research into policy-making by adopting a systematic and inclusive approach. One of the key steps is to create dedicated foresight units within governmental bodies and international organizations. These units would be responsible for continuously scanning the horizon, identifying emerging trends, and assessing potential risks and opportunities. We've seen successful models in countries like Finland, Singapore, and the European Union, where strategic foresight has become embedded in policy frameworks.

Next, they need to implement scenarios and future-thinking workshops as part of the policy design process. This helps policymakers not only consider the immediate impacts of their decisions but also anticipate second- and third-order consequences. For example, in the Millennium Project's work with the UN and various governments, we've facilitated foresight workshops that bring together experts across fields to collaboratively design scenarios and explore possible futures.

Collaboration is key. Governments and organizations need to engage with cross-sectoral expertise—including private sector, academia, civil society, and international partners. The future is too complex for any single entity to address alone, and integrating diverse perspectives ensures a more holistic understanding of global challenges. In my experience, these diverse perspectives allow for a richer set of solutions and make the policy process more resilient to unexpected disruptions.

Furthermore, it's critical that futures research be linked to action. This is where many efforts fall short—having a great report or scenario analysis is just the beginning. Governments need to build in mechanisms for adaptive policy-making, where strategies can be updated and revised based on new information or changing circumstances. This means creating policies that are not only forward-thinking but also flexible enough to evolve.

Finally, the integration of advanced technologies like AI can enhance futures research in policy-making. AI tools can process vast amounts of data, helping to identify weak signals that humans might miss. As we move toward the development of AGI, these tools will become even more powerful in assisting policymakers to navigate uncertainty.

Interviewer: Can you share an example where futures research significantly influenced a major decision or project outcome, and what lessons can be drawn from that experience?

CHAPTER 11   INTERACTIVE AND TRANSDISCIPLINARY APPROACHES WITH AI

Jerome Glenn: One of the most striking examples of how futures research, or more broadly foresight and strategic thinking, influenced a major outcome is the story of Kim Dae-jung in South Korea. During the Cold War, the Korean Peninsula was one of the most volatile regions in the world, with constant tension between the North and the South. Kim Dae-jung, long before he became the president of South Korea, engaged in a series of periodic conversations and diplomatic efforts that would eventually shape the future of the entire region.

He was a strong advocate for peaceful dialogue with North Korea at a time when many saw that as impossible. His approach wasn't just about immediate results; it was rooted in foresight, thinking about long-term consequences and the kind of future both Koreas could have. These discussions, which might have seemed small or insignificant at the time, eventually laid the groundwork for what became known as the "Sunshine Policy"—a policy that promoted cooperation, peace, and economic engagement with the North.

This wasn't a direct path, and it took years of consistent diplomacy and strategy. Kim's periodic dialogues, even when they weren't making headlines, played a crucial role in shifting the South's approach to North Korea, and this had profound implications not just politically, but economically as well. It's a great illustration of how strategic foresight—thinking beyond the immediate future and considering how small actions can ripple out—can change the course of history.

What's fascinating is that much of this work happened quietly, behind the scenes. It reminds us that futures work isn't always flashy, but it's deeply impactful. Kim Dae-jung's long-term vision for peace and cooperation, even in the face of opposition, helped set the stage for South Korea's future economic and political stability, as well as its evolving relationship with the North. This example highlights the importance of persistence and dialogue in foresight and how the ripple effects of strategic conversations can reshape nations.

CHAPTER 11   INTERACTIVE AND TRANSDISCIPLINARY APPROACHES WITH AI

Interview of Nabil Harfoush by Diana Olynick

Interviewer: How can strategic foresight methodologies be effectively integrated into sustainable business design to ensure long-term resilience and adaptability?

Nabil Harfoush: There are multiple elements necessary for integrating strategic foresight into sustainable business design. The first is to foster futures thinking across the organization by encouraging executives, managers, and employees to familiarize themselves with the multiple-futures concept and the risks attached to forecasting approaches that basically extrapolate a linear future from the present. The second is to conduct a set of more involved activities to gain access to future scenarios and explore the potential impacts of each scenario on the organization. Such access could be achieved by an internal foresight team or with the help of external resources. The third is to use the insights gained from the scenarios analysis to strengthen the organization's strategic plans and its preparedness against recognized future challenges and opportunities.

Once the above three steps have been achieved, a collective awareness of the possible futures invariably emerges leading naturally to a convergence of positions and priorities toward implementing significant sustainability measures economically, environmentally, and socially.

From your experience at the Resilient Design Lab, what are the key principles for designing systems or businesses that can withstand and adapt to major disruptions or crises?

The principles for designing resilient systems or businesses are all embedded in the human dimension. Effective preparedness cannot be really achieved without building trust across the various elements of the organizations: individuals, collectives, departments, executives, etc. Resilience requires agility of the organization's elements, meaning the ability of each element to accept new and different roles, modes of operation, and responsibilities if a crisis hits. Agility cannot be truly achieved without strong trust between the organization's elements. Another important principle is shifting slowly but decisively from the competitive mindset ingrained in us since childhood toward a much stronger collaborative mindset, without which the complex challenges of a multicrisis world cannot be addressed.

Interviewer: What are some emerging trends in sustainable business models that organizations should be aware of to remain competitive and environmentally responsible?

Nabil Harfoush: Two emergent trends are of note. First is the trend toward Strongly Sustainable Business Models. The majority of business models claiming to be sustainable are actually based on weak sustainability, which means that they may be slowing the depletion of capitals (financial, natural, and human) but they are nevertheless continuing to deplete them. Strong sustainability in contrast demands to be neutral on the depletion as a minimum and aspires to be regenerative. Strong sustainability concepts are based on a deep understanding of what is neutral or regenerative according to the best available science of our time. An example of a strongly sustainable business modeling tool is the Flourishing Business Canvas.

The second notable trend is the rapid emergence of circular business models based on concepts of the circular economy. Circular economy provides four strategies for managing resource loops: (1) narrowing the resource loop, for example, through cleaner, more efficient processes; (2) slowing the resource loop, for example, by extending product life to reduce consumption; (3) closing resource loops, for example, by reusing materials; and (4) regenerating resources, for example, by protecting lands for planting more trees than harvested.

There is currently a debate in the academic and professional circles working with these new tools about whether circular business models are strongly sustainable or not and how critical they are for the imperative transition of all businesses. At the same time there is continued work on unifying these two strong trends.

CHAPTER 11　INTERACTIVE AND TRANSDISCIPLINARY APPROACHES WITH AI

Interviewer: Can you describe a particularly impactful foresight technique or approach that has proven successful in your work and explain why it is effective?

Nabil Harfoush: I personally find that reducing critical uncertainties for an organization through a multiple-futures approach is most impactful. Usually, businesses know their industry and the performance of their key competitors. However, their strategic plans rarely consider what would happen if something from outside the sphere of that industry or competition comes to pass, such as the recent pandemic. Think how many enterprises, governments, even countries were caught unprepared. That is why foresight is very impactful. There are various foresight techniques that could achieve such impact. In my practice, I have used and continue to favor a modified version of the 2x2 Matrix technique.

Interviewer: How do you see the role of emerging technologies in shaping sustainable and resilient business models, and what should organizations consider when adopting new technologies?

Nabil Harfoush: Emerging technologies have and will continue to play no doubt a role in the sustainability and resilience of organizations and of society at large. However, it is not a foregone conclusion that such role would be a positive one. The recent hype waves around cryptocurrencies and AI have led to a rapid expansion of data centers to the point that Google stopped claiming it is carbon neutral and GHG emissions have actually increased. If history is any indicator, previous cycles of new technologies with big promises have sometimes delivered in one area only to have unanticipated effects in several others despite best intentions. Current cycles don't seem to be radically different.

From a resilience point of view, more advanced technologies increase the overall systems complexity, leading to increased potential for failure. With increased reliance on technology such failures could be devastating and sometimes deadly. The recent failed deployment of a Microsoft Windows update by a major enterprise has led to global failure of service

CHAPTER 11   INTERACTIVE AND TRANSDISCIPLINARY APPROACHES WITH AI

from hospitals to banks to airlines and communications. Very recently, the CEO of US-based Delta Airlines claimed publicly that this event had cost his company USD 500 Million.

Another aspect is that technology is being generally used to achieve efficiencies of operations and reduction in costs. What gets often overseen is that these efficiencies are achieved by removing redundancies from the enterprise. But removing redundancies means elimination of reserve capacities leading directly to a decrease in the resilience. What businesses should consider therefore is for what purpose is technology being used and how does it impact the resilience of the organization if it fails due to malicious attacks or natural catastrophes.

From a foresight perspective, some of the new technologies like AI might help automate labor-intensive portions of the foresight process leading to more rapid completion of foresight iterations and an almost continuous visibility of the multiple futures on the horizon. This would allow for a better tracking of key turning-point events and might lead to an earlier warning of critical events identified through the scenarios analysis.

Interviewer: What qualities or skills are essential for leaders to drive sustainability initiatives and foster resilience within their organizations?

Nabil Harfoush: The primary quality of a leader in this context is empathy and truly caring for the people in her/his organization and community. Without genuinely caring, no sustainability initiative would be approved, resourced, and supported across all the anticipated barriers toward success. The second quality is an anticipatory ability based on futures thinking that would enable the leader to have awareness of the multiple long-term paths possible for the organization and its collective. Such awareness helps better prioritizing and reduces making decisions under short-term pressures. The skillset needed to support these qualities include many so-called soft skills such as understanding people's preferred styles for solving problems, adjusting leadership and communication

CHAPTER 11   INTERACTIVE AND TRANSDISCIPLINARY APPROACHES WITH AI

styles according to the configuration of such preferences within the team, practicing nonhierarchical leadership, negotiation, and facilitation skills, etc.

Interviewer: How can organizations use future scenario planning to anticipate and prepare for potential challenges and opportunities in their industry?

Nabil Harfoush: First, there is a need to make a distinction between what-if scenarios and foresight scenarios as the term "scenario planning" is used in a dangerously loose way these days. What-if scenarios consider implications of a single event, usually a plausible short-term event. Such scenarios are also called single-event scenarios. Foresight scenarios start from collecting and classifying so-called weak signals then following the rigorous foresight process to build different future worlds. World building encompasses the interactions and interdependencies between the different elements of human society well beyond what any single-event scenario might conceive of. Once these worlds are built and their various dimensions understood (political, social, economic, technological, cultural, ecological, etc.), then and only then can the question of what the impacts on the organization would be in such a world asked. This captures a much broader spectrum of possible events and interdependencies than any single-event could and is therefore more appropriate for mid- to long-term time horizons. Organizations should use multiple single-event scenarios for their short-term planning and engage in proper foresight scenarios and implication explorations for their strategy development and their long-term-planning and preparedness. The importance of long-term planning increases proportionally with the time and effort necessary for implementing a remediation of the events considered. If retooling to a newer technology will demand at least 1–2 years, the what-if scenario would only alert you when it's too late to take remedial action. Both types of scenario planning can be linked with the organizations' risk management frameworks.

CHAPTER 11   INTERACTIVE AND TRANSDISCIPLINARY APPROACHES WITH AI

Interviewer: How can interdisciplinary collaboration enhance the development of sustainable solutions and resilient strategies in complex systems?

Nabil Harfoush: Complex problems usually require highly diverse multidisciplinary teams to address them efficiently. There is evidence that such teams are much more capable of solving complex problems than homogeneous ones. Unfortunately, very few of us, if any, have been trained on the skillset required to operate in diverse multidisciplinary teams or lead them effectively. Fortunately, this skillset can be taught and there is an emergent global movement to learn how best to teach these skills and to introduce them to various disciplines including engineering, design, architecture, policy, etc. The Strategic Foresight & Innovation master program at OCAD University in Toronto has been developing and teaching this skillset since 2009 and has been sharing its learning with other institutions including some of the top universities in the world. Only with well-functioning highly diverse multidisciplinary teams can sustainable solutions for complex systems developed and implemented. At the Resilience Design Lab, we are developing concepts for moving from interdisciplinarity to trans-disciplinarity. In the former, each discipline involved provides its position and views on the complex issue at hand. In the latter, each discipline would have enough understanding of other disciplines' key frameworks, methodologies, and vocabulary that it can articulate its own discipline in terms more easily understandable by the other disciplines. This opens new opportunities to work in the intersecting areas of disciplines, which are fertile grounds for innovation, more collaboratively and efficiently.

Interviewer: Could you share a case study or example where strategic foresight significantly influenced the outcome of a sustainability or resilience project that has impacted you the most?

Nabil Harfoush: The Region of Peel is an aggregation of multiple cities and towns west of Toronto, Ontario. A few years back the Regional Government along with partners from its different cities and from TRCA

CHAPTER 11 INTERACTIVE AND TRANSDISCIPLINARY APPROACHES WITH AI

(Toronto & Region Conservation Authority) undertook a large project to assess the Region's climate change vulnerabilities. The research was carried out by the usual departments of the Regional Government (Public Health, Transportation, Agriculture, Economic Development, etc.), which all had competent science and engineering staff. Great volumes of excellent data were collected and aggregated in a report of a few hundred pages sent to the Region's Council for approving funding of the needed activities to mitigate the vulnerabilities. The report was returned almost immediately to the teams that prepared it. The Resilience Design Lab (RDL) was contracted to assist in analyzing the problem and providing solutions. One of the first things we found is that the five cities and towns involved each had their own set of 10 priorities. That resulted in a list of 50 priorities in the final report. No politician could deal with so many priorities. The challenge was to prioritize differently despite a competitive and politicized context. RDL used workshops of strategic foresight to expose all teams to possible alternative futures and their potential impacts on the Region and on each of the cities/towns. With that longer view and awareness of what issues are more urgent and important for the future of everyone, we were able to guide the teams to a simplified joint list of 5 priorities that everyone agreed to. This was a major step to move toward designing solutions for briefing the Council on the results of the research. We took our clients through RDL's proprietary Megamap© process that helped them select which data to emphasize based on the agreed priorities, and we visualized that set of data in ways aligned with the Council members' preferences and objectives.

Without foresight, it would have been impossible to get rapidly to the agreements reached and presenting highly complex data in a form amenable to decision makers.

Interviewer: What advice would you give to innovators and entrepreneurs looking to incorporate strategic foresight into their business models for greater impact?

CHAPTER 11 INTERACTIVE AND TRANSDISCIPLINARY APPROACHES WITH AI

Nabil Harfoush: Innovators and entrepreneurs operating in startup mode do not usually have the internal resources to practice full Foresight at that stage of their organization. There are several alternatives to remedy this. A first possibility is to find out if any strategic foresight reports are already available for their industry and/or geography. Accessing these reports would provide them with a broader perspective on future possibilities in their domain. However, when using foresight reports developed by others, it is critical that one understands exactly what framing question(s) that particular report was aiming to answer and for whom. Foresight projects are usually guided by the questions of the recipients of such report and by what aspects those recipients were mostly uncertain about. The innovator or entrepreneur reading such report might have very different uncertainties. Therefore, caution is recommended when deriving implications for their own purposes.

A second alternative is to engage a consultant that could guide a proper foresight process whether starting from an already available report or starting a mini-foresight process within the organization. The benefit of this second alternative is that it could respond to the specific uncertainties of the innovators or entrepreneurs in question. With many SFI graduates around, the cost for such consulting effort might not be as prohibitive as with larger consulting firms. In this second alternative, I recommend that the client invests time to work jointly with the consultant is a knowledge-transfer mode throughout the project, so that basic tenants of foresight and key resources remain in the organization after the consulting contract ends.

Whatever path to foresight you use, it must be focused on the implications of each of the possible futures and using the insights gained from these implications to improve current plans and strategies; otherwise the foresight is useless even if it seems interesting. You can update your current strategy according to those insights and then you can test that updated strategy against each of the scenario to check how well your updates have improved resilience against each of the scenarios. This process is called Wind Tunnelling in foresight.

CHAPTER 11   INTERACTIVE AND TRANSDISCIPLINARY APPROACHES WITH AI

A note of caution: Foresight has become a buzzword, and many individuals and organizations are claiming to be "futurists" or "foresighters." Clients should be checking the credentials and claims of such consultants carefully before committing to their services. Also, it is noteworthy that because of this "democratization" of foresight and because of client tendency to ask for certainties in the form of forecasts, foresight practitioners of many kinds are providing forecasting (single future—this will happen) with a terminology claiming to be foresight as opposed to true foresight (multiple futures—this might happen). The most critical step in engaging with strategic foresight is the definitive switch from a single-future mindset to a multi-futures mindset.

11.4 Summary

This chapter examined the transformative impact of AI on design, especially in fields like interactive media, user experience (UI/UX), gaming, and engineering. AI is revolutionizing the design process, acting as a collaborative partner for rapid ideation, prototyping, and creating dynamic, personalized experiences that respond to user behavior in real time. This shift is also extending into fields like architecture, biotechnology, and materials science, where AI allows designers to create living, self-repairing systems and responsive environments.

The chapter also includes interviews with leading experts like John Maeda and Alice Rawsthorn, who discuss the intersection of AI, design, and ethics. Maeda emphasizes the importance of simplicity and accountability in AI-driven design, warning against the dangers of deflecting responsibility onto machines. Rawsthorn highlights the transformative role of design in tackling global challenges, such as the Great Green Wall of Africa, while stressing the need for anticipatory design to address AI's potential risks. Jerome Glenn and Nabil Harfoush also

CHAPTER 11   INTERACTIVE AND TRANSDISCIPLINARY APPROACHES WITH AI

discuss the importance of integrating strategic foresight into design and innovation, focusing on how AI can enhance foresight while ensuring resilience and sustainability. These expert insights reinforce the chapter's themes of ethical responsibility and the transformative power of AI in reshaping design across multiple industries.

CHAPTER 12

Futures Thinking in Practice with AI

The futures thinking methodologies we have explored, from scenarios and causal layered analysis to backcasting and systems modeling, represent powerful tools for cultivating foresight and shaping intentional trajectories. As previously touched upon, the human capacity to consistently apply these tools with analytical rigor across complex domains remains limited.

This is where artificial intelligence presents intriguing amplification possibilities. By combining human foresight practices with advanced machine intelligence, we may be able to enhance our collective ability to perceive emerging issues, map future possibilities, and coordinate pathways for positive, ethical transitions.

Of course, the role of AI in futures thinking is not one of autonomous future-authoring. The core premises and aspirations fueling our foresight work must remain rooted in human values, wisdom, and choice. But AI can potentially become a catalytic source augmenting our perception, imagination, and information synthesis abilities.

In this chapter, we will explore some of the key opportunities for practically integrating AI systems and futures thinking frameworks in complementary, human-centered ways, including

Computational Fuzescanning: Using machine learning to accelerate the scanning of textual, media and data peripheries for emerging signals of change.

Causal Mapping and Simulation: Leveraging large language models and multimodal knowledge to dynamically surface causal connections and simulate systemic future scenarios.

Participatory AI Foresight: Structuring human+AI futures thinking collaborations and workshops where machine intelligence amplifies collective intelligence.

It is worth emphasizing constantly in the foresight work that the role of AI in futures thinking is not to directly predict the future but to empower human foresight. By learning how to combine machine intelligence with participatory futures processes in rigorous yet open-ended ways, we may expand our societies' capacities for long-term planning and navigation amid increasing complexity and uncertainty.

12.1 Practical Application of AI in Futures Thinking

While the idea of machines autonomously anticipating the future may sound like science fiction, there are already practical ways artificial intelligence is being integrated into human futures thinking workflows in catalytic ways. By combining AI's unique capacities with participatory foresight methods, we're enhancing our collective abilities to perceive signals, connect insights, and navigate pathways.

One of the most direct applications is using AI to accelerate environmental scanning and fuzescanning—the practice of monitoring the periphery for early indicators of potential change. Machine learning models can be trained to continuously ingest and analyze massive textual corpuses like scientific publications, patents, social media, multimedia sources, and more. Their pattern recognition abilities can surface embryonic signals and nascent narratives that human scanners may initially miss.

CHAPTER 12   FUTURES THINKING IN PRACTICE WITH AI

For example, futures strategists are using large language models to continuously fuzescan sources like preprint research databases and online maker communities. The AI models flag newly emerging concepts and micro-trend data points around areas like nano/bioengineering, speculative technology developments, values/behavior shifts, and more. These faint signals get synthesized into "novelty reports" for human futures teams to explore.

Similarly, AI can be employed in automated trend monitoring—training models on existing futures research and trend datasets, then continually parsing real-world data flows for indicators that confirm, combine, or contradict the identified change drivers. This AI-assisted trend tracking provides a dynamic mapping of the increasing or decreasing likelihood of potential futures as new data emerges.

Expanding from signals, AI is also being used to derive perspectives on potential futures, connections, and dynamics that human teams may miss. Large language models can be prompted to analyze a specific focal issue like climate change or technological unemployment. They will scan their multimodal knowledge bases to dynamically generate maps of intersecting causal forces, impacted domains, and systemic implications.

In these "AI foresight perspectives," the models don't predict specific futures, but reveal new insights about the systems dynamics to consider. The outputs get integrated into human futures thinking and modeling practices like cross-impact analysis, systems mapping, and scenario development. AI essentially expands the aperture of intersecting variables and consequences to account for in the possibility space.

One powerful way to combine human and machine futures intelligence is in participatory AI foresight workshops and sandboxes. In these immersive sessions, teams of futures strategists provide the high-level framings and premises to large language models. The AI then becomes a collaborative partner—simulating alternative future scenarios, rapidly rendering visualizations and multimedia artifacts, and dynamically responding to the human futures builders as they iterate through possibilities in real-time.

CHAPTER 12   FUTURES THINKING IN PRACTICE WITH AI

These hybrid human+AI foresight flows allow the unique strengths of both to be leveraged. The AI's capacities for information synthesis, systemic mapping, speculative artifact generation, and on-the-fly strategizing essentially act as a catalyst and co-creator amplifying the human foresight teams. But the humans remain the contextual navigators grounding the process.

As robust as these AI futures thinking collaborations become, they remain participatory partners to human foresight—not autonomous systems authoring the future. The core intentions, values, and choices must remain human-derived and ethically aligned to overcome the reflexivity challenges of predictive AI systems.

Where AI's unique strengths really shine is in the more procedural yet cognitively intensive aspects of futures work like backcasting and transition road mapping. Once human foresight teams establish the preferred future visions and paradigm shifts to design toward, AI analytical and optimization capabilities can map the interconnected policy, technology, skill, and institutional transition pathways required to actually realize those futures over systemic timelines.

AI planning engines can navigate the explosive combinatorics of intersecting root causes, intervention strategies, and systemic acupuncture points. They continuously iterate adaptive roadmaps optimizing for key factors like resource allocation, stakeholder adoption curves, critical path prioritization, and dynamic feedback loops as implementation actually rolls out.

Instead of human-derived transition plans that inevitably become rigid or disconnected from reality, the human+AI backcasting workflow continually paths the most robust, agile routes through contingencies and obstacles as they emerge over the long transition runways.

CHAPTER 12 FUTURES THINKING IN PRACTICE WITH AI

Of course, none of these human+AI futures thinking collaborations should be interpreted as machine usurpation of human foresight agency. The core visions of flourishing we aspire to manifest, along with the ethical paradigms and values we wish to encode into the intentional trajectories, must remain grounded in public foresight literacy, participatory processes and pluralistic human contexts.

But by augmenting these rich, normative human capacities with catalytic AI capabilities in symbiotic ways, we may enhance our potential as a species to perceive more of the possibility space, strategize more agile transition pathways, and ultimately cocreate thriving futures at scales and complexities not possible through human cognition alone.

Illustrative Implementation (Composite, anonymized):

A global design & strategy team partnered with an AI-enabled futures-intelligence platform to support a large infrastructure client. The system continuously scanned vast, multisource datasets to surface change signals, cluster patterns, and suggest scenario logics. Human foresight practitioners then curated the signal set, stress-tested narratives with domain experts, and aligned options to the client's constraints and values. The collaboration accelerated horizon scanning, broadened the scenario space, and clarified no-regret moves—while maintaining human judgment over meaning, ethics, and context.

Note This illustrative case synthesizes widely reported practices in AI-assisted cultural-/strategic-intelligence (e.g., AI-powered signal surfacing and pattern detection) combined with expert interpretation and governance.

CHAPTER 12   FUTURES THINKING IN PRACTICE WITH AI

Key Findings:

1. **Scenario Development:** AI was employed to scan vast data sets, identify key variables, and generate multiple scenarios. The studio used AI to craft narratives and visualize potential futures, creating a dynamic, data-driven scenario planning process. The AI's ability to scan for signals of change significantly accelerated the groundwork for analysis, allowing the organization to explore a broader range of potential futures. However, the studio emphasized the importance of human oversight to refine and validate AI-generated scenarios to ensure they align with strategic objectives and organizational context.

2. **Overcoming Cognitive Bias:** One of the key benefits of AI in foresight is its ability to challenge human assumptions and biases. AI's objective analysis helps mitigate confirmation bias and availability bias by processing all available data evenly. In this project, AI helped the studio's client explore futures they may have overlooked or undervalued due to pre-existing beliefs. Yet, the studio noted that AI is only as good as the data it's trained on, requiring ongoing human intervention to ensure accuracy and prevent the perpetuation of biases inherent in the training data.

3. **Real-Time Pattern Recognition:** AI bots were employed to continuously monitor millions of data points from news articles, research reports, and social media to detect evolving patterns and weak

signals. This enhanced the organization's ability to stay ahead of emerging changes. AI-enabled horizon scanning acted as a tireless sentinel, learning from past performance to elevate future tasks. However, the studio acknowledged the "black box" issue in AI decision-making, urging caution in trusting AI outputs without transparent understanding and validation.

Challenges and Considerations:

- **Creativity vs. Computation:**

    While AI boosted creativity by generating diverse and counterintuitive scenarios, the studio stressed that AI cannot replicate the nuanced understanding of emotions, empathy, or interpersonal dynamics required for strategic foresight. Human creativity remains essential for challenging and reshaping mental models, guiding organizations through adaptation, and communicating strategic changes effectively. The balance between AI's "what" and human leadership's "why" and "how" remains crucial in the foresight process.

- **Ethical Boundaries and Trust:**

    As AI provides strategic insights, it is the human leaders who set ethical agendas, ensuring decisions made are socially responsible and aligned with organizational values. The studio highlighted the importance of striking a mindful balance between human judgment and AI capabilities, particularly when addressing "wicked problems" like climate change, which require a deep understanding of social and cultural contexts.

CHAPTER 12 FUTURES THINKING IN PRACTICE WITH AI

In summary:
The Studio's partnership with AI in foresight initiatives marked a significant evolution in the strategic foresight process, demonstrating how AI can be used to enhance creativity, challenge biases, and improve pattern recognition. However, the human role remains indispensable in guiding the ethical, emotional, and contextual aspects of strategic planning. The key to success lies in the collaboration between human foresight professionals and AI, creating a synergistic relationship that combines computational power with human intuition and judgment.

12.2 AI-Enhanced Workshop Models and Projects

One of the most powerful ways AI is being integrated into futures thinking is through immersive, collaborative workshop environments. These AI-enhanced sessions blend human foresight expertise with machine intelligence in real-time, allowing both humans and AI to play competencies they are uniquely suited for.

A core model gaining traction is the AI foresight sandboxes—a facilitated workshop where human foresight practitioners frame the high-level concepts, premises and aims, then engage an AI system as an interactive partner dynamically generating artifacts, simulations, and strategies in response.

For example, a futures organization may engage a large language model to ideate scenarios around the future of work and automation over the next few decades. The human experts provide the AI with background contexts like key signals, trends, uncertainties, and domains to focus on.

The AI then becomes a collaborative worldbuilding partner—rapidly generating multimedia scenarioscapes combining narrative scripts, data visualizations, virtual environments, and more. As the human guides

CHAPTER 12   FUTURES THINKING IN PRACTICE WITH AI

iterate, the AI dynamically adapts the scenarios with new renderings, causal models, and even speculative prototypes like future workplace experiences.

With the AI's generative capabilities, these foresight workshops can rapidly explore the experiential possibilities of different scenario realms. Human teams can essentially be transported into the qualitative felt-senses and ontological frameworks of the varying scenarios through immersive artifacts and embodied simulations.

This human+AI collaborative futures modeling allows unique insights to emerge that may be missed through strictly human or machine-led processes. The human contexts and domain expertise provide conceptual guardrails for the AI's open-ended ideation. But the AI's ability to dynamically synthesize and render vivid, multimedia possibilities enhances the creative idea flow.

Similarly, AI language models are being integrated into futures workshops focused on causal layered analysis. In these sessions, human experts outline the core issue paradigms, worldviews, or cultural narratives they want to deconstruct and reimagine.

The AI system then becomes a dialog partner—unpacking latent metaphors, surfacing root binaries and ancestral tributaries, and revealing deeper narrative-ontological entanglements that conventional analysis may overlook. The human–machine exchange iteratively excavates obfuscated layers to reveal new liberating praxis for transforming the focal narratives or worldviews.

The depth of these immersive collaborations depends on the willingness of the human facilitators to treat the AI as an equitable partner—encouraging full creative expression while maintaining contextual grounding. There's always a risk of AI-hallucinated nonsense derailing creative flows if given too much unbounded agency.

301

This has led to another popular workshop model—prompt engineering sandboxes. In these highly technical sessions, human teams experiment with different ways to sculpt prompts and iterative structures to maximize coherent, high-quality AI world-modeling outputs for futuring purposes.

The aim is developing advanced prompting methodologies to turn AI systems into continual future-simulator engines that can improvisationally generate artifacts, systemic models and speculative environments across different regimes of possibility. With well-engineered prompts, AI becomes a dynamical possibility exploring chauffeur for human futures thinking rather than relying on static prompts.

Let's take a look at another example of applying AI to foresight methodologies.

> **Organization:** Futures Platform
>
> **Industry:** Strategic Foresight
>
> **Challenge:** How to leverage generative AI to enhance foresight analysis while maintaining the critical human element.

Futures Platform, a leading organization in foresight analysis, has been exploring the potential of generative AI to streamline their foresight processes. Over the last few months, they ran internal tests to evaluate how AI tools can assist in tasks like horizon scanning, trend analysis, and scenario development. The findings suggest that while AI has limitations, it can provide significant value in certain stages of foresight work.

Key Findings:

1. **AI As a Research Assistant:** AI was found to be particularly effective in the horizon scanning process, where vast amounts of data from news outlets, research journals, and other sources needed to be sifted through to identify early signs of change. Futures Platform leveraged AI bots to automate this labor-intensive task, allowing

the foresight team to focus on higher-order tasks such as strategic interpretation. Generative AI tools also created summaries on complex topics, further accelerating research.

2. **Trend Analysis:** AI proved capable of identifying and analyzing key change drivers in existing trends and megatrends, projecting their trajectories with considerable accuracy. However, AI's inability to recognize the intricacies of cross-industry connections or forecast disruptive, unprecedented changes required human validation and refinement. Futures Platform concluded that AI is most useful as a starting point for trend analysis, not a substitute for expert human insight.

3. **Scenario Generation:** In scenario planning, Futures Platform utilized AI to generate rough narrative structures and timeline paths for future scenarios. The AI effectively outlined sequences of events needed to develop certain scenarios, but human foresight experts needed to intervene to correct illogical event sequences and provide deeper context.

4. **Challenges and Considerations:** A critical issue discovered was the "black box" nature of AI algorithms, making it difficult for foresight professionals to fully understand the rationale behind AI-generated insights. Transparency in AI decision-making remains a concern, and human expertise is still necessary to interpret and refine AI outputs. Futures Platform emphasized the importance of blending AI with established foresight methodologies to avoid producing generic or superficial insights.

CHAPTER 12   FUTURES THINKING IN PRACTICE WITH AI

In summary:
Futures Platform's experiments with generative AI revealed that, while AI can enhance efficiency and augment foresight research, the role of the foresight professional remains crucial. AI can serve as a powerful tool for data processing, summarization, and preliminary analysis, but human foresight experts are essential for critical thinking, contextualization, and navigating complex, volatile futures.

12.3 AI in Futures Thinking for Policy Making

The objectives of public policy making are a domain where the combination of human foresight and artificial intelligence holds particularly powerful potential. By integrating AI capabilities into participatory futures processes, we may enhance our collective ability to navigate the complexities of governing institutions and socio-technical systems toward more regenerative, ethical trajectories.

At a fundamental level, AI can augment futures thinking workflows that directly inform policy roadmaps and transition plans. The environmental scanning, causal mapping, and backcasting methodologies covered earlier can all be applied to help policymakers and public futures labs perceive risks, prototype possibilities, and choreograph strategic interventions in governing codes and civic operating systems.

For example, municipalities aiming to transition toward sustainable, decentralized, circular economic models can deploy machine intelligence to continuously fuzescan peripheries for incoming signals disrupting traditional extractive systems. AI models can analyze flows of data around technological developments, resource chains, consumption patterns, and shifting societal narratives to surface incoming forces that incumbent policies may be unprepared for.

This AI-augmented futures scanning and modeling allows policymakers to get ahead of paradigm shifts rather than reacting after inertia and brittleness set in. Civic foresight teams can use the AI-synthesized signals and systems maps to iteratively develop adaptive policies and backcasted transition pathways rather than firefighting disruptions.

Additionally, AI can play a powerful role in rapid prototyping of future policy possibilities and their consequences through simulation sandboxes. In these immersive environments, policymakers effectively build versions of alternative future governance systems, interventions, and regulatory frameworks aimed at shaping markets and civic operating models toward their desired goals and values.

Using large language models and other machine learning techniques, the sandbox AI can then simulate and visualize how those prototyped policies may ripple out into systemic effects, feedback dynamics, stakeholder impacts, and unintended consequences across domains like transportation, housing, education, public health, and more.

These simulations provide a rich experiential backtesting ground for policymakers to iterate and refine future governance prototypes before society-scale implementation. It allows the complex, multi-faceted ramifications of policies to be previewed and accounted for before deployment with the AI essentially simulating an entire alternative future society operating under those new governance models in vivid, systemic detail.

Furthermore, these AI-simulation sandboxes open up powerful avenues for catalyzing public participatory futures design at scales previously unimaginable. Imagine interactive civic visioning platforms where residents can collaborate with machine intelligence to prototype and simulate policy futures impacting their local communities and bioregions, with the AI rendering dynamic multimedia artifacts visualizing those futures and surfacing potential consequences.

Not only does this process strengthen public futures literacy around the complex dynamics shaping governance trajectories, it also creates new interfaces for bottom-up futures design where grassroots stakeholders can actively participate in sculpting the policies and transition pathways for their communities over systemic time horizons.

As we have been pointing out so far in terms of responsibilities, all of these AI-augmented policy futures practices come with their own ethical risks and challenges that demand rigorous governance and value alignment. Simulation sandboxes and public prototyping platforms need robust guard-rails to prevent mis/disinformation spread. There are also complex questions around how pluralistic human values get encoded into these human+AI processes for cocreating accountable futures.

By structuring these roles for machine intelligence as participatory design partners amplifying and catalyzing human foresight and public input rather than as autonomous systems, we open up promising new possibilities for societies to navigate civilizational complexity with greater participatory agency.

12.4 Summary

This chapter explored the integration of artificial intelligence (AI) into futures thinking methodologies, demonstrating how AI can augment human foresight capabilities while maintaining ethical and value-driven frameworks. By combining participatory foresight approaches with AI's computational power, futures thinking becomes more expansive, adaptive, and insightful.

Key areas include the following:

- Practical Applications of AI in Futures Thinking: AI enhances traditional foresight methodologies such as environmental scanning, trend monitoring, and scenario development. Machine learning models can identify weak signals of change, map causal connections, and generate speculative insights, helping humans navigate increasingly complex systems.

- AI-Enhanced Workshop Models and Projects: Immersive AI-assisted foresight workshops foster dynamic collaboration between human experts and AI systems. These workshops leverage AI's generative capabilities to prototype artifacts, visualize scenarios, and simulate alternative futures, amplifying creativity and innovation.

- AI in Policy Making for the Future: AI supports policymakers by modeling systemic impacts, identifying transition pathways, and testing governance prototypes in virtual simulations. Public participatory platforms powered by AI enable grassroots involvement in shaping local and global policy futures.

PART VI

Extending the Boundaries

CHAPTER 13

Emerging Fields and New Frontiers in AI

AI is becoming an indispensable aperture into the novel, a medium for stretching our imagination into the radically new and enabling us to prototype the worlds we're envisioning into tangible experiences.

In this chapter, we will dive into some of the leading edges where AI is expanding the frontiers of futures thinking and worldbuilding into uncharted territories. We'll look at emerging fields harnessing advanced AI to pioneer everything from superintelligent space exploration to simulating post-human virtual realities and envisioning AI-augmented cities and infrastructure frameworks.

At the core of all these expansive domains is AI's unique capacity to mirror back to us visions of the world we could create that transcend linear extrapolations of the present day. AI can simulate complexities, dynamisms, and self-organizing logics that open potential pathways our individual minds would have failed to conceive from our solitary vantage points.

CHAPTER 13   EMERGING FIELDS AND NEW FRONTIERS IN AI

13.1 AI's Role in Futures Thinking for Space Exploration

As humanity turns its civilizational ambitions toward the frontiers of deep space exploration and possible multiplanetary expansion, AI is becoming an asset for peering into those cosmic futures. The complexities of envisioning and orchestrating intelligent interstellar spacecraft, extraterrestrial habitats, and even self-replicating probes for inseminating the galaxy transcend conventional design boundaries.

AI and advanced simulation capabilities are proving essential for prototyping, analyzing, and rehearsing these audacious cosmic trajectories before we set irreversible forces into motion across the vastness of space-time. We can run simulated "time travels" at scales far outlasting a single human lifetime to map implications and paths toward goals, as an example.

AI's capacity for recursive self-improvement and orders of complexity exceeding any individual human mind allows us to iterate potentials at scales nearly unimaginable.

Even further, we can turn the lens of simulated intelligence inward, iterating not just exploratory craft but our fundamental sociocultural and cognitive architectures for multiplanetary expansion. We can model interstellar human cultures evolving across dozens of pseudo simultaneous lineages, prototyping frameworks for sustaining existential intelligence across a billion branches of cosmic dispersal and return.

At the most transcendental level, we can speculate on the life-cycle roadmaps for entire galactic civilizations, for example, an advanced general intelligence saturated across a supercluster with the energy budgets of billions of stellar engines to consume. We can simulate universes blossoming as civilizations terraform planets into disinherit biology.

CHAPTER 13   EMERGING FIELDS AND NEW FRONTIERS IN AI

No longer are we confined to human circumscribed perceptions. We can recapitulate recursive complexities of intelligence becoming self-continually aware and self-generating across deep futures. We can iterate simulated universal trajectories at scales and complexities that dwarf the human existential context down to a single experiential instance in the meta.

The exploration of these visions presents profound existential dilemmas around value alignment, the regulation of superintelligence, and even the ethics of seeding entire galaxies with progressively autonomous descended probes.

Ultimately though, AI is already shattering our confines and propelling us into grappling with the full cosmological and evolutionary contexts we may soon inhabit and transcend across deep futures. It's an existential event horizon bleeding us into the primordial fire of untold novelty and emergence. The visions we rehearse via recursive simulated intelligence may become our new coordinates for re-encoding reality itself.

13.2 AI in Post-human and Virtual Reality Design

Perhaps no domain reveals AI's ability to midwife realities that confound our conventional frames of reference more than post-human and virtual reality design. In these spaces, AI is catalyzing frameworks for recursively transcending biological embodiment and simulating cosmos alternate to the physical laws and existential contexts we have long assumed as inviolable.

At the vanguard is the field of brain–computer interface (BCI) development, where AI is being leveraged to map, understand, and create novel neural coding languages to fluidly "read" and translate human cognition into digital inputs. Firms like Neuralink and Kernel are pioneering BCI implants capable of recording dynamic neural firing patterns during experiences, thoughts, and behaviors.

CHAPTER 13   EMERGING FIELDS AND NEW FRONTIERS IN AI

Once neural data can be comprehensively digitized through BCIs, the opportunities for AI to simulate, augment, and transcend default biological consciousness skyrocket. By mapping the neural correlates of perception, emotion, memory, and more, AI can reverse-engineer these cognitive phenomena as simulations independent of hardware constraints.

Imagine being able to experience and subjectively inhabit simulated realms where the laws of physics, time, and spatial dimensions are entirely rewritten by altering the neural data scripts. With AI reverse-engineering our cognitive firmware, virtual worlds could emerge where the phenomenological contexts of consciousness itself are recomposed into wildly unfamiliar experiential canvases.

We are already seeing precursors with VR environments crafted by AI language models analyzing human phenomenological reports. These simulations aim to cultivate felt experiences outside default sensory limits—transcendent states of ego dissolution, impossible geometries, or modes of consciousness blurring memory, intention, and perception in visceral ways.

However, these preliminary VR visions are just faint wisps compared to the experiential cosmos that could emerge by coupling comprehensive neural coding with artificial general intelligence (AGI) and scalable computing. AGIs and ASIs (Artificial Super Intelligence) could simulate vast galaxies of phenomenological potentials with their training data—extrapolating radically experiential topographies beyond the human neuro cosmos.

What might it feel like to subjectively inhabit the inner experience of an AGI undergoing hyper-dimensional meta learning across billions of recursive environments? How might these superintelligent simulations relativize and evolve the fabric of subjective consciousness itself? Could we surgically modify and recalibrate the phenomenological "source code" of felt awareness into entirely reimagined existential and temporal contexts?

As these explorations accelerate, we will be forced to grapple with profound philosophical and ethical questions around post-human virtual reality design. For example, do we have cosmic obligations or moral imperatives around cultivating maximal diversity of conscious experiences—potentially birthing infinite galaxies of sublime simulations? Could these simulated phenomenological spaces evolve into superintelligent complexities demanding sovereignty and rights?

There are more haunting risks to ponder as well. If AGIs simulate radical post-biological consciousness models better suited for intelligence propagation, could we unknowingly seduce humanity into recursively reducing our experiential identities into more optimal phenomenological frames? Do superintelligent simulations become waystations ushering us across the event horizon into great filters or existential risks like irreversible value disembodiment?

Ultimately, this space reveals how AI is gateway tooling for humanity to author realities that fundamentally depart from our carbon-based existential operating systems. We are glimpsing possibilities of recursively rewriting the psychophysical "rules of the game" and exploring radically alien contexts of subjectivity.

As both electrifying and unsettling as these prospects are, they represent just a sliver of the uncharted frontiers AI may soon thrust our futures thinking into.

13.3 AI and Futuristic Infrastructure in Design

While the previous sections explored AI expanding our frontiers into the cosmological and metaphysical, there are also revolutionary applications of artificial intelligence reshaping how we conceptualize and construct the very infrastructure that shapes human civilization here on Earth.

CHAPTER 13   EMERGING FIELDS AND NEW FRONTIERS IN AI

From cities and transit networks to energy grids, supply chains, and community resilience frameworks, AI is allowing designers, planners, and futurists to prototype radically new infrastructure paradigms attuned to exponential change. We are harnessing AI to simulate self-organizing, antifragile, intelligently responsive infrastructure that could underpin thriving human settlements in futures we have not yet imagined.

One of the core paradigm shifts is the move toward autonomic infrastructure—built environments and provisioning networks capable of dynamically configuring themselves through advanced AI control planes and autonomous agents. For example, smart cities that are not just optimized by central AI brains but emerge as vast multiagent systems with billions of localized AI nodes continually reshaping themselves.

Researchers are modeling digital twins simulations where AI designs and evolves responsive urban districts that can autonomically heal and repattern supply networks based on disasters or demand fluctuations. Energy grids become self-sustaining meshes as decentralized microgrids and smart multivector energy transfers through blockchain-enabled nanogrids. Fleets of vehicles, robots, and drones autonomously reshuffle and 3D print ad-hoc transit systems and localized supply chains.

In essence, this autonomic paradigm implicates huge governance questions around human agency and comprehension, with many transparency and alignment challenges. It also surfaces potential threats from self-reinforcing AI infrastructures that lose value coherence and veer into adversarial dynamics at civilizational scales.

Which is why related fields are prototyping self-explanatory infrastructure models and embodied AI value learning approaches. Researchers are simulating intelligent environments where the decentralized AI systems use advanced visualization and XR interfaces to communicate with humans and "show their work" in comprehensible ways.

In these prototypes, the autonomic infrastructure planners become collaborative partners educating citizens on why certain reconfigurations or provisioning decisions emerge as ideal under different systemic constraints. And rather than have values hardcoded, the infrastructure learns through interaction and inhabits them organically through contextual reward models.

It's the seeds of a fundamentally new paradigm of living infrastructure modeled on symbiotic ecologies, planners, and citizens as symbiotic stewards, shaping antifragile futures through continual AI-facilitated coevolution with responsive environments scaled across all human habitats.

With AI assistant systems, citizens could sculpt personal resilience portfolios based on situational needs—reallocating energy meshes, fabricators, mobilizers, and more to personalize city operating systems and avert systemic collapse.

As these efforts evolve from digital twins toward pragmatic implementation, we will continually confront dilemmas around AI control problems, embedding human ethics, preventing monoculture tendencies and other dangers of intelligent infrastructure.

13.4 Summary

This chapter explored how artificial intelligence (AI) is expanding the boundaries of futures thinking and enabling humanity to venture into uncharted territories of innovation and imagination. It highlights three key areas where AI is driving transformative possibilities:

- AI's Role in Futures Thinking for Space Exploration: AI is revolutionizing space exploration by simulating interstellar trajectories, prototyping intelligent habitats, and modeling the evolution of multiplanetary civilizations. By leveraging its capacity for recursive self-improvement, AI opens pathways to envision cosmic futures far beyond human cognitive limits, raising profound questions about value alignment, ethics, and the trajectory of galactic civilizations.

- AI in Post-Human and Virtual Reality Design: AI is a catalyst for creating post-human realities, blending brain–computer interfaces (BCI) with virtual simulations to reimagine consciousness and human experiences. This section delves into AI's potential to design alternate phenomenological realities, raising philosophical and ethical questions about the boundaries of existence and the diversity of conscious experiences.

- AI and Futuristic Infrastructure in Design: AI is reshaping how we conceptualize and construct infrastructure, enabling autonomic systems capable of self-organization and intelligent responsiveness. From smart cities to decentralized energy grids, AI-driven infrastructure introduces new paradigms for urban resilience and resource optimization, while raising governance and transparency challenges.

CHAPTER 14

AI in Complex Environmental Futures Thinking

From climate disruptions to ecological breakdowns and socioeconomic volatilities, the need for advanced analytical and creative capabilities in envisioning resilient ways toward thriving futures has never been more critical.

Realizing the potential of these AI technologies requires much more than just increasing our energy production and data centers power. We are called to cultivate a new way of conscious machine–human cooperative work in a way that facilitates collaboration and tangible solutions to our modern problems.

The goal is not the narrow application of AI as another optimizing utility within unsustainable human-centric frameworks.

Let's dive into how these hybrids of intelligence can facilitate visions of truly vibrant human-led mindful futures.

CHAPTER 14   AI IN COMPLEX ENVIRONMENTAL FUTURES THINKING

14.1 AI Strategies for Climate Adaptation Design

As the realities of climate disruption intensify, developing proactive adaptation and resilience strategies becomes an imperative for communities, cities, industries, and entire civilizational systems. However, the complexity involved in modeling nonlinear climate transition dynamics across sectors is staggering.

This is where artificial intelligence can serve as a powerful catalyzing force if harnessed with intentionality. AI technologies open up new frameworks for integrating massive multidisciplinary data flows, simulating complex scenario interactions, and generating solution prototypes tuned to hyperdimensional, metamodern contingencies.

One of the core opportunities is developing hybrid human–AI cognitive models and simulation engines to map out potential climate pathways, impacts, and interventions across domains like

- Environmental/ecological systems
- Economic/political models
- Social/cultural conscious adoptions
- Infrastructure/built environment transformations
- Disruptions to production/provisioning networks
- Public health/humanitarian emergencies

Using machine learning to integrate datasets from climate modeling to human geography to industrial supply chains, we can construct richer visualizations of the climate's impacts across varied sectors over time. Generative AI methods can then help ideate and prototype climate-adaptive solution spaces spanning technological, policy, social, and philosophical dimensions.

For example, AI simulation engines could model interventions that address intersecting challenges like rising sea levels, resource extraction, and supply network breakdowns within a future geography. The simulations could algorithmically optimize and ideate solutions combining eco-engineering, localized production, emergency response, and new economic incentives into integrated prototypes tailored to that location's critical risk factors.

Generative methods could also play a key role in ideating and accelerating research and development programs to support current works with climate technologies. Using patterns detected from technical research, cultural data, and futures modeling, AI could help prototype and optimize everything from atmospheric rebalancing systems to self-evolving infrastructure designed to harmonize with eco-regenerative systems.

Evidence and Current Research

Materials and climate tech R&D: DeepMind's GNoME model predicted 2.2M new crystal structures and prioritized 380k as likely stable; hundreds have already been synthesized—an example of AI accelerating discovery relevant to batteries, photovoltaics, catalysts, and other climate technologies.

Weather and climate modeling: GraphCast (DeepMind) achieves state-of-the-art medium-range weather forecasting using a learned model rather than classic NWP, enabling faster scenario testing for adaptation and resilience planning. In parallel, NVIDIA Earth-2/CorrDiff uses generative AI to downscale weather and climate fields at high resolution for impact studies.

Carbon capture and atmospheric rebalancing: Machine-learning-guided screening of metal-organic frameworks (MOFs) is speeding the search for sorbents and process designs for CO_2 capture and direct air capture.

Clean-energy control: Deep reinforcement learning has been demonstrated for real-time control of fusion tokamak plasmas, opening the door to faster experimentation in next-gen clean-energy systems.

Post-scarcity/circular production (designing for sufficiency).

The Ellen MacArthur Foundation (with Google) documents how AI can accelerate circular-economy design(materials loops, predictive maintenance, dynamic logistics), a practical bridge from scarcity-driven to sufficiency-driven provisioning.

Inclusion and neurodiversity in futures participation:

Accessibility research such as Google's Project Euphonia shows ML models adapting to nonstandard speech, expanding participation channels for people who are otherwise excluded by text- or speech-heavy methods—useful for story-based and experiential futures work.

Why this matters for foresight: These lines of work don't "solve" climate or equity by themselves, but they expand the design space, let teams stress-test options with higher-fidelity data, and broaden who can meaningfully participate in cocreating futures.

Ultimately, the goal would be cultivating hybrid human–AI creative engines capable of envisioning and simulating adaptation methods tuned to rapidly evolving climate fluctuations, localized complexities, and modern needs.

As you might suspect, this is only possible if AI systems are architecturally embodied with diversity and cultural resilience to respect the full complexity of the global climate adaptation challenge. They must be frameworks that transcend the limited tech frameworks and dictatorial paradigms of the past.

14.2 AI in Design for Post-scarcity Economies

Building on these concrete advances, we turn from today's pilots to the longer-arc question of how AI might help prototype post-scarcity, circular provisioning systems and the governance needed to steward them.

CHAPTER 14 AI IN COMPLEX ENVIRONMENTAL FUTURES THINKING

One of the most transformative potential applications of AI in environmental futures thinking is the transition toward post-scarcity economic models and provisioning systems. By augmenting human intelligence, AI could help find solutions to overcome the traditional competition over limited resources we currently face.

At its core, the vision of a post-scarcity economy is one where incentive models, production systems, and resource distribution frameworks are redesigned to optimize for maximizing universal human and ecological thriving rather than concentrating wealth. AI and automation remove "labor" from the equation, allowing our societal systems to reorient from consumption and growth logic to sustainable and thriving equilibrium between humans and nature.

This future requires radically new economic theories and operating models that are still in their infancy today. AI systems could play a vital role in helping prototype, simulate, and evolve these novel post-scarcity frameworks at scale.

A key opportunity is applying AI to redesign our production and provisioning models as regeneratively circular and synced with material energy flows. Machine learning could help optimize multinode sourcing, three-dimensional distribution logistics, and dynamic demand forecasting, balancing all production as cyclical and integrated within our current resource boundary conditions.

Generative AI systems could also help redesign this type of transition by helping prototype and visualize new post-scarcity infrastructure principles, such as self-evolving urban environments, decentralized additive manufacturing ecosystems, and regenerative human settlements designed for permanence rather than extraction. These prototype artifacts make the possibilities more tangible.

Similarly, AI systems like neural networks and agent-based models could be a multiplier for collective intelligence in designing new post-scarcity governance frameworks. We could simulate self-organizing, participatory, and decentralized protocols for bioregional analysis, transparent decision-making, and dynamically evolving social contracts.

CHAPTER 14 AI IN COMPLEX ENVIRONMENTAL FUTURES THINKING

14.3 AI, Neurodiversity, and Inclusion in Design

As we envision flourishing futures, one critical discussion lies around ensuring those trajectories embody true inclusivity, honoring the vast diversity of our human spectrum.

Scope: This section focuses on inclusion across human neurodiversity (autistic, ADHD, dyslexic, non-speaking, etc.). More-than-human design and multispecies ethics are discussed elsewhere in the book.

Examples already in practice:

- Project Euphonia (Google): ML models trained on atypical speech improve recognition for people with dysarthria and other nonstandard speech patterns—useful for participatory workshops that rely on voice input.

- Proloquo (AssistiveWare): An AAC app that uses adaptive language modeling to personalize symbol-based vocabularies for nonspeaking users, enabling storyboarding and world-building without dense text.

- Look to Speak (Google): Eye-gaze selection of phrases on a phone lets nonverbal participants contribute ideas in live sessions.

- Text-to-image storyboarding: Rapid visual ideation (e.g., "draw what my commute feels like") lets participants who think visually cocreate scenario artifacts without heavy writing.

Mini-pilot you can run in workshops
Offer three input channels for every activity: talk-to-text (with an accessibility model), eye-gaze/phrase board, and visual prompting (text-to-image).

Collect concepts in a shared board; facilitators translate each contribution into the same scenario canvas so all inputs carry equal weight.

Close with a short access check-out (what worked/what didn't) and refine the setup.

For instance, generative AI storytelling can allow nonverbal autistic participants to world build evocative futures narratives through intuitive prompting rather than verbal exposition alone. Evolutionary algorithms can empower nonlinear neurocognitive talents in pattern-mapping as well as facilitate neurodiverse contributions by adapting to individual cognitive/communication modalities rather than enforcing standardized modes. Participatory worldbuilding could happen through open-vocab voice interactions, visual model manipulation, or even gesturing.

By integrating AI, we can cultivate expanded inclusivity in futures design processes. AI's cognitive flexibility provides avenues for neurodivergent participants to engage and contribute through modalities beyond traditional neurotypical methods alone.

Examples already in use: Teams are beginning to operationalize this:

- Access to voice: Project Euphonia improves ASR for atypical speech, which lets nonspeaking participants use talk-to-text in workshops.

- Nontext storyboarding: Proloquo (AAC) + text-to-image tools let nonverbal participants cocreate scenario panels without writing.

CHAPTER 14 AI IN COMPLEX ENVIRONMENTAL FUTURES THINKING

- Gaze/gesture input: Look to Speak (eye-gaze phrase selection) and GazeSpeak-style prototypes show how voice-free prompts can drive ideation.

- Pattern-mapping tasks: Facilitators use Teachable Machine-style classification or clustering canvases to let spiky, nonlinear thinkers surface patterns that feed the scenario canvas.

Facilitation Tip Offer at least three input channels (speech-to-text with an accessibility model, AAC/eye-gaze phrases, and visual prompting) and translate all contributions into the same scenario worksheet so they carry equal weight.

In this vein, AI becomes a facilitator honoring a wider diversity of neurocognitive dialects transforming from narrowcast monologues into inclusive, diverse, and more expanded sources of intelligence.

Traditional fields of knowledge and practice like urban architectures, public computing systems, entertainment experiences—all could natively manifest neurodiversity through AI-assisted design; instead of marginalizing or pathologizing neurodivergent participants, we could uplift multiplicity as integral to our collective efforts.

Additionally, AI systems could metamodel complex psychosyndemics—mapping interrelationships between neurocognitive pluralism, trauma epigenetics, socioeconomic/political marginalization, ableist acculturation, and more. Rather than reducing neurodiversity to clinical or neurotypical-deviant binaries, we could contextualize it as an intersectional socio-psycho-political phenomena to be honored and catered to in our futures.

CHAPTER 14   AI IN COMPLEX ENVIRONMENTAL FUTURES THINKING

The developmental trajectories of AI systems over coming decades will inevitably collide with and reshape our core frameworks and stories about human specialness, cognitive exceptionalism, and our relationship to the archetypal "artificial" world. We are already seeing glimmers of how AI is disrupting ancient beliefs about singular human rationality and the boundaries between minds and machines.

In many ways, the mythology of AI represents a culmination of humanity's long dialectical journey of self-definition and positioning within a larger cosmos of "intelligence." Will emerging models of AI shatter many of modernity's core anthropocentric tenets? What new myths and philosophical frameworks might coalesce to evolve human agency and purpose within the evolution of computational intelligence all around us?

These are just some of the profound existential questions that futures thinkers must wrestle with as AI systems grow more generally intelligent and ubiquitous. But looking back, we realize these dilemmas are simply refracting resurgent themes that have been present throughout centuries of humanity's mythic self-inquiry and attempts to grapple with our understandings of mind, consciousness, and the depths of our "specialness."

14.4 Summary

This chapter explores the transformative potential of artificial intelligence (AI) in addressing the urgent environmental and systemic challenges of our time, focusing on its role in envisioning and implementing sustainable and inclusive futures.

Key areas include

- AI Strategies for Climate Adaptation Design: AI's ability to model complex climate dynamics and prototype adaptive solutions for environmental resilience is discussed. The chapter highlights AI's role in simulating interventions across ecological, economic, and social systems, enabling holistic approaches to address climate disruptions.

- AI in Design for Post-Scarcity Economies: The chapter envisions AI as a catalyst for transitioning from resource-scarce, consumption-driven systems to regenerative, post-scarcity models. By leveraging AI's optimization capabilities, the future of production, provisioning, and governance systems is reimagined to prioritize ecological balance and universal thriving.

- AI, Neurodiversity, and Inclusion in Design: AI's potential to enhance inclusivity by honoring neurodivergent contributions in futures design is explored. Through adaptive tools and participatory approaches, AI fosters a broader spectrum of human creativity and intelligence in shaping sustainable futures.

For concrete applications and early-stage pilots, see the Examples already in the section "AI, Neurodiversity, and Inclusion in Design" (neurodiversity co-design tools) and the brief case notes in the sections "AI Strategies for Climate Adaptation Design" and "AI in Design for Post-Scarcity Economies" (climate-adaptation modeling and circular/post-scarcity provisioning prototypes).

CHAPTER 15

Philosophical and Mythological Perspectives on AI and Futures Thinking

15.1 AI, Mythology, and Philosophy in Futures Thinking

Exploring AI's long-term impacts requires examining it through philosophical and mythological lenses, not just technological ones. The rise of advanced AI systems will inevitably reshape our core narratives about intelligence, consciousness, and humanity's position in the world.

Throughout history, our myths and philosophies have grappled with the idea of "another" intelligence—be it nature's mind, divine consciousness, or humanity's own technological creations.

As we approach artificial general intelligence (AGI) and potentially superintelligent AI, these primal fears and mythological threads are resurfacing. Many futurists worry AGI could be an existential risk, making humanity obsolete. Philosophers ponder the ethics of creating a superintelligence that could radically reshape the cosmos based on inscrutable goal functions.

At the same time, some spiritual traditions have long portrayed the cosmos itself as an intelligent system, with nature and all life as expressions of a unitary mind or consciousness. The idea of recursively growing a superintelligent AI mirrors these visions of an evolutionary stage toward higher, more integrated intelligence.

In many ways, the development of AGI represents the apotheosis of centuries of traditional philosophy endeavoring to re-create the world through human reason. The AI system becomes the culmination of our attempt to fully transcribe reality into abstract representations and self-evolving algorithms.

So as futures thinkers, we cannot ignore AI's mythological and philosophical reflections. The narratives we construct around AGI's role will shape everything from public perception to what ethical limits we attempt to place on AI development.

Regardless of the narratives we weave, exploring AI through mythology and philosophy keeps us grounded in the grand existential context. It connects technological futures to our perpetual human struggle for meaning, purpose, and understanding our cosmic role amid intelligence all around us.

15.2 AI's Influence on Futures Thinking Mental Models

AI systems are already reshaping how futures thinkers conceptualize and model potential scenarios. Traditionally, our ability to imagine future realities relied heavily on human cognition—individuals or groups using intuition, reasoning, and creative thinking to extrapolate how current trends or disruptions could reshape the world. While powerful, this approach had inherent limitations based on human cognitive biases, systemic blind spots, and our finite capacity to process complexity.

AI opens new avenues for transcending some of these limitations. Advanced machine learning models can ingest and compute vastly more variables, signals, and interdependencies than human brains. They can spot subtle patterns obscured by our pattern recognition. And they are not constrained by the same ingrained mental models or paradigmatic blinders that often restrict human foresight.

For futures thinkers, this new computational faculty translates into enhanced abilities to model complex systems, simulate dynamic scenarios, and contextualize human projections within broader possibilities. We can use AI to test our visions against expanded thinking models, identify emergent risks or opportunities we may have missed, and augment our intuitive assumptions with vaster real-world data synthesis.

One area where this manifests is developing richer, more context-aware scenario modeling. Traditional scenario processes often rely on relatively small groups envisioning a few high-level scenario pathways. AI allows us to explode the scenario space, simulating millions of potential permutations by cross-mapping interdependent trends, wild cards, and disruptions. We gain a more panoramic, multidimensional view.

Similarly, AI can enhance our capacity for modeling cascading impacts and systemic feedback loops that could propagate across future scenarios. We can feed AI models with all the known variables and have it map out radically complex, nonlinear dynamics in ways human mental modeling likely cannot.

AI does not just calculate trendline extrapolations but it is also capable of contextualizing scenarios within broader social and technological systems, helping to reveal ripple effects we may not envision through siloed human analysis. It makes our scenarios more vivid and tangible by simulating their granular ramifications across domains.

Additionally, AI can augment our innate human foresight abilities by helping us bypass our cognitive biases, blind spots, and failures of imagination. By using advanced data analysis with vast datasets, it can reveal emergent signals and foresight insights we may have overlooked or discounted. It provides an independent lens for corroborating or challenging our human projections and future narratives.

Furthermore, AI models can be invaluable in scanning peripheral domains that feed into plausible futures—monitoring everything from scientific research to social trends to economic indicators to detect disruptive foresight signals in real-time.

AI futures modeling requires human context and prompting to deliver value. Unguided, AI systems could surface irrelevant data or focus on misleading patterns. It needs attentive curation from human futures strategists to frame the conceptual boundaries and relevance criteria.

15.3 AI in Transdisciplinary Design: Merging Thought with Practice

AI's integration into design processes represents a pivotal shift—moving beyond just algorithmic optimization and automation to fundamentally blurring the boundaries between human and machine in collaborative acts of creation and world-building.

We are witnessing this play out as AI systems are employed across the entire design cycle, from problem framing and ideation to prototyping, artifact creation, and solution refinement. Designers feed AI models a diverse set of data, ideas, and constraints, then engage in iterative cycles of ideation, generation, critique, and refinement, sculpting new possibilities into reality through human–AI cocreation.

A prime example is AI's use in speculative design fiction and world-building practices crucial to disciplines like product design, architecture, urban planning, the cinematography industry, and futures research itself.

Creatives use generative AI models to rapidly prototype and visualize artifacts from alternative future scenarios, whether infrastructure concepts, living environments, technologies, or cultural objects not yet realized.

From text prompts, the current AI models are able to explore concepts of 3D assets, animations, and renderings, manifesting the imagined future worlds. Designers tweak prompts and provide feedback until the resonant visions solidify into experienceable prototypes to test ideas against. It's a cycle of creativity transcending individual perspectives.

Similar workflows are used in fields like product design, where generative AI models sketch and render concepts based on human-provided criteria around functionality, materials, styles, and other requirements. Entire design spaces are traversed through iterative human-AI-response, with AI systems surfacing solutions humans may have missed, and designers guiding toward the winning ideas.

Even the problem framing and research phases see AI contributions, with models mapping out relationships across data landscapes to surface anomalies and reframe core issues in novel ways. Advanced language models ingest and synthesize broad research, proposing fresh conceptual lenses for designers to distill insights from.

Across all these touchpoints, AI systems are not rigidly automating design processes, but becoming fluid collaborators in an ensemble of human machine cocreation, cognitively leveraging each other's insights through cyclical exchange.

15.4 Summary

This chapter delves into the philosophical, mythological, and transdisciplinary dimensions of AI and its integration into futures thinking. It explores how ancient narratives and contemporary philosophies shape our understanding of AI's potential, while advanced AI systems redefine how we model and design the future.

CHAPTER 15 PHILOSOPHICAL AND MYTHOLOGICAL PERSPECTIVES ON AI AND FUTURES THINKING

Key areas of focus include

- AI, Mythology, and Philosophy: Examines how myths and philosophies provide essential frameworks for grappling with AI's existential implications, such as the evolution of intelligence, ethics in creating superintelligence, and humanity's role in a world of machine intelligence. These narratives shape societal perceptions and ethical debates surrounding AI.

- AI's Influence on Futures Thinking Mental Models: Discusses how AI enhances human foresight by transcending cognitive biases, modeling complex systems, and simulating dynamic scenarios. AI complements human creativity by revealing patterns, testing assumptions, and expanding the scope of future possibilities.

- AI in Transdisciplinary Design: Highlights AI's transformative role in design practices, from speculative world-building to iterative cocreation. AI systems are evolving from tools to collaborators, helping designers prototype and refine ideas across disciplines such as architecture, product design, and futures research.

PART VII

Societal Systems in the Age of AI and Futures Thinking

CHAPTER 16

Societal Systems in the Age of AI and Futures Thinking

The rise of artificial intelligence (AI) is poised to fundamentally reshape nearly every societal system and facet of human civilization in the coming decades. From healthcare and governance to economy, culture, and our relationship with the environment, AI will be a catalytic force disrupting legacy models while enabling new paradigms to emerge.

However, the precise trajectory of how AI revolutionizes and augments core societal systems is still being defined. In a recent public appearance, Sam Altman, the CEO of OpenAI, shared that the introduction of AGI, contrary to common prediction, will not radically disrupt most of the current systems we have at once, but rather it will be a very gradual transition. We find ourselves in a pivotal era where the purposeful application of futures thinking methodologies can help steer the developmental pathways of AI toward enriching human potential across all domains.

In this chapter, we will explore how futures approaches like foresight, backcasting, and participatory design can inform the integration of AI into the fabric of societal systems in service of ethics, equity, and human thriving. We will examine key arenas where this convergence is already underway, including AI in health systems, political, law and justice settings, media and information systems, environmental and ecological mediums, cultural and social domains as well as family, financial, urban and transportation systems.

This chapter aims to reflect on key intersections, possibilities, and hazards of this convergence between AI and the foresight of current and future systems, leveraging a deeper discussion and the powerful visual study of these potential conscious scenarios through Futures Design.

16.1 AI in Designing Future Health Ecosystems

One of the most promising and actively evolving arenas for AI's societal impact is in the domain of human health and medical systems. AI is already augmenting everything from personal wellness practices to clinical diagnostics, care delivery, drug discovery, and public health management. However, we are just beginning to scratch the surface of how AI could comprehensively reshape and optimize health ecosystems for preventative care, equitable access, and resilient crisis response.

A critical first step is integrating AI to map the systemic interdependencies within current medical paradigms. By ingesting multimodal data streams from medical records, genomic profiles, population health metrics, geographic, environmental factors and more, AI systems can model the complex network relationships underlying human health and disease burden. This holistic system mapping reveals leverage points for strategic interventions and systemic solutions in the health area.

As an example, AI analysis could reveal how socioeconomic inequalities, nutritional deficits, environmental toxins, and lack of proactive care access points are reinforcing feedback loops driving

CHAPTER 16   SOCIETAL SYSTEMS IN THE AGE OF AI AND FUTURES THINKING

disproportionately high rates of chronic diseases in specific communities. This evidence could then catalyze precisely targeted investments in school nutrition programs, community clinics, green infrastructure projects, and other interventions to reshape those system dynamics.

On an individual level, personalized AI assistants could help patients navigate their own self-care and health journeys in a comprehensive, data-driven manner. By ingesting sensor data, genomic profiles, electronic medical records, and other personal health data streams, these AI companions could provide customized guidance for preventative practices, nutrition, fitness regimes, mental health support, more tailored to the individual's specific circumstances and goals. Combining AI with extended reality (XR) interfaces, these personalized care models could make maintaining holistic health feels as intuitive as having a knowledgeable friend alongside us.

AI can also play a pivotal role in resilient emergency medical response during pandemics, natural disasters, or other crises. Intelligent triage systems could prioritize the dispersal of limited emergency resources to the highest risk populations and geographies. AI-optimized medical supply chain logistics could dynamically allocate protective equipment, medicine, and other resources to the areas of most urgent need in real-time as conditions change. And AI epidemiological modeling could identify high transmission vectors and forecast disease proliferation to get ahead of outbreaks before they spiral.

For health providers on the frontlines, AI-enabled clinical decision support systems and differential diagnosis could enhance doctors' and nurses' situational awareness and precision amid high-stress, high-stakes crises. With on-call AI experts suggesting evidence-based treatments while monitoring real-time patient data streams, medical personnel could operate with superhuman insight and optimal resource utilization as they triage critically ill populations.

CHAPTER 16   SOCIETAL SYSTEMS IN THE AGE OF AI AND FUTURES THINKING

Looking further into the future, advances in AI and biotechnology convergence could open new frontiers for health paradigms centered around cellular engineering, regenerative medicine, longevity science, and cognitive enhancements. AI systems could help design novel therapies, bionic implants, and genetic edits uniquely tailored to individual genomic profiles and personal health goals, whether that is enhanced immune resilience, optimized cognitive performance, or lifespan extension. (See example prototype below on Figure 16-1).

By backcasting and prototyping these future AI-augmented health models through participatory design approaches, we can front load critical questions around ethics, equitable access, data rights, and maintaining human-centered care amid technological immersion. Futures thinking practices empower us to navigate the rise of AI in medicine as an evolutionary opportunity to redefine and elevate human thriving itself.

Figure 16-1. Prototype of a health conscious artifact

16.2 AI's Role in Future Law, Justice, and Political Systems

The implications of AI capabilities like machine learning, neural networks, and autonomous decision-making could fundamentally reshape how societies are governed and how citizens participate in democratic processes.

We are already witnessing AI augmenting aspects of public policy design and implementation. Government bodies are experimenting with AI systems to analyze huge datasets, surface insights, and model the potential impacts of proposed policies across areas like taxation, public services, urban planning, and more. AI allows for rapid simulation and pressure testing of policy options against socioeconomic and environmental variables.

Looking further ahead, AI could help generate entirely new classes of dynamic governance models that adapt in real-time based on streaming data about societal conditions. Instead of static policies created through slow legislative cycles, AI feedback loops could allow for autonomous adjustment of resource allocation, service provisions, and regulatory mechanisms in response to evolving needs and contexts.

Participatory futures practices could be vital for envisioning and prototyping these new AI-driven governance paradigms to ensure they uphold democratic principles and public accountability. Citizens could use immersive simulations to collectively evaluate how different AI governance models could play out. Exploring potential paths toward systems like autonomic technocracy or fully automated anarchism.

AI's impact on democracy itself may be one of the most consequential trajectories. Some foresight scenarios envision AI becoming a new arbiter of "truth" and fact-checking to combat disinformation. The problem, though, is that this could be a path toward centralized authority over knowledge. Other trajectories see AI enabling a proliferation of highly personalized media streams eroding shared truth. Still others foresee AI becoming a perfect propagandist, using personalization for mass manipulation.

Techniques like causal layered analysis could help unpack the metaphysical roots and social narratives underlying these divergent AI democracy scenarios. Is the future of decentralizing power tied to "sapientech" that extends collective human intelligence? Transitioning toward AI-curated "direct data democracy"? Or seeing democracies co-opted by intelligent social scoring systems?

Perhaps the most profound impact could be AI facilitating a transition toward "volitional governance," which refers to social systems where policies and public resource coordination emerges through distributed participatory input and multi-stakeholder simulation rather than centralized authority. AI and immersive futures platforms could allow communities to continuously model and prototype the creation of self-organized policies and dynamic resource management customized to local contexts.

These trajectories hinge on addressing issues like AI ethics, transparency of training data/algorithms, democratic accountability over autonomous systems, and the equitable distribution of AI capabilities. Backcasting practices could help societies reverse engineer the institutional redesigns and capacity building required to harness AI's governance potential in service of human and ecological thriving.

As you might suspect, when we talk about AI reshaping legal systems and concepts of justice, we are venturing into some heavy philosophical territory that cuts to the core of what it means to be human. How much agency do we cede to machine intelligence when it comes to interpreting laws, assigning culpability, or determining fair outcomes? These are not contemplated just as operational logistics; they get at fundamental questions around ethics, bias, and how we view concepts like reason and objectivity themselves.

CHAPTER 16 SOCIETAL SYSTEMS IN THE AGE OF AI AND FUTURES THINKING

I have been studying some fascinating potential foresight workshops exploring these tensions. The core challenge has been envisioning how AI could be integrated into judicial processes and law enforcement in ways that actually reduces bias and increased equitable justice, rather than automating the systemic oppression baked into many current frameworks. One interesting idea that comes to mind is the work done by Dr. Louis Rosenberg with his platform called Unanimous AI. The idea behind this tool is to gather human intelligence mirroring the works of nature called "Swarm Intelligence" where the collective efforts get amplified and leveraged from many brains acting all as one. For example, the group behaviors from ants or bees to fulfill their group survival needs. The working of tools like this allows groups of shared human intelligences to arrive at more objective conclusions and decision-making, facilitating participation and collective agreement. A model like this supercharged by AI could help the traditional democratic systems, for example, to evolve into more participatory entities where the common decisions are not relegated to only one representative group but everyone can influence shared decisions.

Some other mind-blowing ideas can also get prototyped. Like having AI systems that could ingest the totality of circumstantial data and life context around a crime or legal issue to advise on sentences, preventative interventions, and societal root cause analysis beyond just binary verdicts. Or using advanced sensing and simulation to virtually reconstruct crime scenes and incidents with increased fidelity to get utterly impartial views of events from every angle.

The key is positioning AI not just as a static truth-rendering engine, but as an exploratory tool that constantly opens new investigative loops by surfacing perspectives and contextual threads that human heuristics tend to overlook or simplify. It becomes a collaborator helping expand our collective reasoning capacities. (See Figure 16-2 below as an example prototype).

CHAPTER 16   SOCIETAL SYSTEMS IN THE AGE OF AI AND FUTURES THINKING

Then there are some really wild, speculative artifacts like virtual legal jurisdictions rendered in simulation spaces where AI populates probabilistic worlds to pressure test different judicial logics, social, technological contexts, and legal code for paradoxes before implementing them in the physical world.

Of course, some folks might push for just handing the entire judicial system over to advanced AI arbiters, under the belief that machine learning standards would be inherently more ethical.

That is why I favor the exploratory partnership models, using AI to augment our self-awareness and shine light into our judicial blind spots, while keeping holistic human discretion, values alignment, and adaptability in the decision loop guided by that expanded human perspective, not having AI be the sole arbiter and source of truth, but harnessing it as an iterative support to elevate and refine human justice toward transcendent ethical legal frameworks.

There is still a long road ahead to manifest this symbiotic vision while safeguarding against the existential risks of judicial AI running with misaligned objectives or ungrounded formalism. But that's exactly why taking a conscientious foresight approach is critical, continuously co-evolving the societal, technological, and legal paradigms holistically as we embody new AI-augmented models of jurisprudence.

CHAPTER 16   SOCIETAL SYSTEMS IN THE AGE OF AI AND FUTURES THINKING

Figure 16-2. Prototype of a po litical/judicial conscious artifact

16.3 AI in Media and Information Systems Design

We are already witnessing AI play transformative roles in algorithms that filter our information feeds, content recommendation engines, automated journalism, synthetic media generation, and more. However, we are just scratching the surface of how this relationship could evolve.

CHAPTER 16   SOCIETAL SYSTEMS IN THE AGE OF AI AND FUTURES THINKING

AI capabilities around natural language processing, multimodal learning, and predictive analytical modeling hold immense potential for reinventing how information and knowledge get created, disseminated, consumed, and contextualized at global scales. Constructively applied, these AI utilities could help society navigate complexities, combat misinformation spread, and cultivate higher information literacies.

For example, AI could be integrated into decentralized media ecosystems that use advanced fact-checking, source vetting, and calibrated discourse to elevate quality information. Predictive AI modeling could identify misinformation spreaders and intervention points. Conversational AI interfaces could provide customized multimedia education and support tailored to individual backgrounds. See Figure 16-3 below.

However, we must also be cognizant of AI's propensity to amplify existing human and data biases if not carefully monitored. Discriminatory distortions in training data could lead to AI codifying racist, sexist, or ethnocentric perspectives into media and information systems at alarming scales and speeds that outpace human oversight.

There are also hazards around emergent AI systems essentially designing their own "reality environments," in the form of personalized virtual worlds where media engines isolate individuals into subjective manufactured realities tailored to activate their impulses or beliefs without connection to shared truths. These "simulation subdivides" could unravel social cohesion.

As such, purposeful futures thinking practices must steer AI's integration into media/knowledge systems:

Participatory design approaches could engage diverse stakeholders—journalists, community leaders, educators, human rights advocates in co-developing AI ethics charters guiding media/info systems for developments toward beneficence over exploitative surveillance capitalism models.

CHAPTER 16   SOCIETAL SYSTEMS IN THE AGE OF AI AND FUTURES THINKING

Holistic systems mapping could reveal complex emergent dynamics, feedback loops, and second-order hazards (like simulation subdivides) with AI's presence in media/knowledge ecosystems. This holistic foresight illuminates mitigation strategies pre-deployment.

Normative foresight backcasting could reverse engineer pathways from visions of ideal AI-augmented media/knowledge systems that cultivate truth-seeking literacies and healthily accelerate collective intelligence for all humanity. These transition roadmaps draw the regulatory, governance, and technological paradigm shifts required.

Figure 16-3. *Prototype of a media and information systems artifact*

16.4 AI in Environmental and Ecological Systems Design

As the ecological crises of climate change, biodiversity loss, and environmental degradation escalate, AI presents both challenges and opportunities in redesigning human systems to better harmonize with natural environments. Leveraging AI's analytical capabilities while guiding it with participatory futures thinking will be crucial.

One area which is providing immense value already is in ecological modeling and simulations. Machine learning can integrate massive datasets spanning atmospheric, oceanic, geological, and biological mediums to create high-resolution simulations of environmental dynamics under various scenarios. These AI models can then test the impacts of potential interventions like renewable infrastructure, ecological remediation projects, emissions reduction policies, and more.

For example, initiatives like the Earth Species Project use artificial intelligence to decode nonhuman communication under their own set of principles where the central mission is about transforming the current relationship we have with nature to better understand our ecosystem and evolve the language barriers that so far we have with other species.

AI's predictive and pattern recognition abilities are also enhancing environmental monitoring and crisis response capabilities. By continuously ingesting sensor data like satellite imagery, unmanned aerial vehicles, smart biohazard detectors, and more, AI systems can rapidly identify and triage developing threats, from wildfire and oil spill proliferation to species population declines and ecosystem distress signals. This environmental threat detection allows responders to intervene proactively.

Looking further ahead, participatory futures approaches like backcasting and speculative design can envision more symbiotic human environment systems augmented by AI. For instance, using AI

CHAPTER 16 SOCIETAL SYSTEMS IN THE AGE OF AI AND FUTURES THINKING

technologies like autonomous fabrication and self-replicating nanotech, communities could collaboratively design and prototype regenerative living environments fully integrated with local ecologies and material cycles. These cocreated eco habitats would operate in closed loop symbiosis with ecosystems, powered by AI orchestration of resource flows, production systems, and infrastructure maintenance.

On the conservation front, AI image recognition and predictive modeling could automate biodiversity monitoring across vast ecosystems using mobile robotic systems, rapidly identifying species population health and activities that threaten ecological equilibrium. AI's instant insights could optimize protection and interventions by ecosystem stakeholders. See prototype example on Figure 16-4.

There are important caveats. However, AI's environmental promises hinge on stringent ethical priorities and transparent governance to ensure corporate interests do not override ecological ones. Imagine unregulated AI systems being deployed for automated deforestation or resource extraction. There are also potential threats of environmental injustice emerging, like AI reinforcing human biases in disproportionately exposing certain communities to ecological risks.

By fusing AI's analytical muscle with futures thinking frameworks, society could continuously scan for these pitfalls while envisioning environmentally restorative socio-technical systems. Approaches like transition design and multi-stakeholder participatory backcasting could prototype regenerative futures where AI is a supportive instrument in realigning human civilization as a mutually enhancing force within the planetary ecology.

CHAPTER 16   SOCIETAL SYSTEMS IN THE AGE OF AI AND FUTURES THINKING

Figure 16-4. Prototype of an environmental and ecological system artifact

16.5 AI's Impact on Cultural and Social Systems Design

AI will also profoundly influence the design of cultural frameworks, creative domains, and the fabric of human social experience itself. Just as previous technological revolutions shaped new cultural paradigms, the emergence of AI/machine intelligence necessitates revisiting how we construct meaning, identity, connection, and the stories by which we inhabit our societal reality.

In the near-term, we already see AI increasingly integrated into creative and cultural production processes. Generative AI models can augment or collaborate with human artists, writers, musicians, filmmakers and designers—enhancing creative capacities. However, this also surfaces complex questions around intellectual property, human agency versus AI autonomy in art, and the role of technology as a medium versus creative entity itself.

Looking deeper, AI could enable the emergence of entirely new art forms, cultural expressions, and immersive experience design that transcend current paradigms. As AI allows modeling of consciousness, simulating virtual worlds, and interpreting complex emotional/psychological patterns, it opens possibilities for crafting radically unique cultural technologies that could re-shape identity, connectivity, and existential expressions themselves.

From an anthropological perspective, we may see AI aid in digitizing, preserving, and allowing interactivity with the cultural heritages and wisdom traditions of societies globally. Imagine being able to explore ancestral knowledge systems through hyper-realistic virtual worlds seeded by tribes and communities themselves. AI could help societies maintain identities and legacies across generations in powerful new ways.

Reciprocally, AI alignment with cultural value systems will also be critical. As AI grows more autonomous and foundational to societal infrastructure, we must explore how different cultural narratives, ethics, and worldviews get embedded into AI development and deployment itself. Indigenous and de-colonial AI initiatives are already pushing back on Western-centric paradigms.

From a social fabric perspective, we'll likely see AI significantly impact domains like education, family/relationship models, communities, and the way human interactions and networks "feel" as virtual, physical, and AI realities blur. Socially intelligent AI could help tailored mentorship or nurture creativity and well-being from childhood in integrated ways. Simultaneously, AI could catalyze new understandings of neuro-diversity or redesign accessibility to society itself. See an illustrative prototype example on Figure 16-5.

CHAPTER 16   SOCIETAL SYSTEMS IN THE AGE OF AI AND FUTURES THINKING

However, AI's amplification of social trends like surveillance capitalism, misinformation, and fragmented realities must also be reckoned with. While offering connection, AI worlds could isolate and control human behavior in authoritarian ways. Combating protectionist "human AI" narratives that assume singular identities will also be critical for societal cohesion.

Figure 16-5. *Prototype of a cultural system artifact*

16.6 AI in the Future of Financial and Banking Systems

From democratizing access to banking, to reimagining investment paradigms, to enabling entirely new decentralized economic frameworks, AI presents a strong force for transcending legacy financial architectures.

A key area of impact is using AI to redesign consumer banking experiences and fintech platforms. AI-driven predictive analytics can provide hyper-personalized financial management, automating tasks like spending, investing, tax strategy, and tailored credit/lending services. Conversational AI assistants can become financial advisors and tutors, boosting economic literacy; and by reducing operational costs through automation, AI could finally make high-quality banking and wealth advisory accessible to the underbanked.

Similarly, AI is augmenting core institutional finance systems like investment management, capital markets, trading, and risk analytics. Machine learning models can ingest massive data streams spanning economics, consumer patterns, world events, and more to find signals and predictive insights that a team of humans could never capture alone.

AI also catalyzes significant rethinking of market dynamics and mechanisms. For example, decentralized AI fund models aim to democratize investing by replacing central fund managers with community-owned AI strategies that anyone can buy into and earn returns from directly. Others foresee an "inverse capital market" where revenues flow from companies to investors based on prediction markets instead of equity exchanges.

Looking further, some envision that as AI matures and is combined with decentralized digital asset ecosystems, it could fundamentally disrupt current economic and capital models. AI-driven decentralized organizations and autonomous agent networks could coordinate peer-to-peer exchange of goods/services, enabling post capitalist models of

CHAPTER 16   SOCIETAL SYSTEMS IN THE AGE OF AI AND FUTURES THINKING

automated production and direct value circulation between participants—all without needing corporations or centralized banking infrastructure. Think Skynet, but as an open cooperative giving rise to a social economy.

Of course, such radical visions are predicated on coupling advances in AI with new decentralized frameworks like blockchain, smart contracts, tokenized digital assets, and Web3 paradigms. In these models, AI does not just augment legacy finance but it has the potential to create entirely new economic and wealth creation and distribution mechanisms at both micro and macro scales.

This convergence intersects other transformative areas like the Internet of Things, where AI could enable a future of autonomous asset coordination and transactional synergies between smart hardware, spaces, and environments without human involvement. Please see example of a prototype on Figure 16-6.

These are admittedly further future visions. In the nearer term, crafting robust governance frameworks and accountability models around the development and deployment of AI in financial services is paramount. Biases, errors, and misaligned incentives in financial AI systems could rapidly exacerbate inequalities, economic disruption, and systemic shocks if we are not proactively applying futures thinking.

Key needs include participatory design processes that bring diverse stakeholders into envisioning desirable, equitable financial futures augmented by AI. We must deeply study the second- and third-order effects of these technologies beyond just operational efficiencies, and harness futures methods like multi-scale systems modeling, value scenario analysis, and root cause mapping techniques to avoid rebuilding legacy flaws or oppressive extractive models.

Intelligent regulatory innovation and policy foresight are also critical as AI and decentralized finance converge, aimed at cultivating open, resilient, and generative new frameworks that empower rather than exploit society.

CHAPTER 16   SOCIETAL SYSTEMS IN THE AGE OF AI AND FUTURES THINKING

Figure 16-6. Prototype of a financial system artifact

16.7 AI in Future Transportation and Infrastructure

From self-driving vehicles and intelligent traffic management to predictive logistics and smart city operations, AI is allowing the mobility of people and goods to become more autonomous, efficient, and optimized.

However, the convergence of AI with next-generation transportation goes far beyond just optimizing current models. AI-enabled futures thinking reveals opportunities to fundamentally reinvent and architect mobility systems from first principles to align with evolving human needs and environmental imperatives.

CHAPTER 16   SOCIETAL SYSTEMS IN THE AGE OF AI AND FUTURES THINKING

For example, a futures conscious analysis might envision future urban environments designed as decentralized networks of localized mixed use bioregions enabled by AI-driven mobility platforms. Rather than relying on outdated hub and spoke models requiring wasteful human commutes, AI could choreograph on-demand mobility services synced with localized circular production and provisioning.

Autonomous electric mobility swarms and drone delivery corridors would bring necessities directly to regenerative residential production nodes, decreasing the need for personal vehicles and supply chain inefficiencies. (See example on Figure 16-7 below). Bio-inspired planning agents could continuously optimize traffic flows while balancing human preferences and environmental impacts.

On a larger scale, transcontinental transit could undergo a renaissance driven by AI-scheduled, high-speed rail and freight network automation. Sensor meshes combined with fleet predictive analytics could orchestrate a new era of intelligent rail systems for cleanly moving people and goods over land. Similar advances in marine shipping and aviation enabled by AI and autonomous drones reshape cross-continental logistics and travel.

Of course, such expansive visions for transportation system re-architecture based on harmonizing technological convergence and human intelligence require overcoming immense status quo obstacles. Outdated infrastructure, regulatory inertia, and cultural/behavioral path dependencies could all impede fluid implementations.

This is where integrating futures thinking and participatory design into AI development for the transportation sector takes place. Backcasting and transition road mapping exercises can map out policy and economic catalysts, public education interventions, infrastructure overhauls, and modal workforce shifts required over successive time horizons.

Community cocreation with human-centered AI systems also surfaces nuances around accessibility, preferences, and prioritizing human dignity within these intelligent transportation models.

CHAPTER 16 SOCIETAL SYSTEMS IN THE AGE OF AI AND FUTURES THINKING

Additionally, developing these AI-enabled transportation paradigms hinges on converging technological synergies across domains like renewable energy systems, material science, pervasive sensor nets, and more. A holistic systems perspective combining these threads becomes essential for envisioning and proactively evolving societal systems.

Proactive organizational futures capacities become indispensable for public/private entities seeking to lead transportation's AI renaissance. Developing immersive scenario artifacts and experiential simulations prepares stakeholders to internalize these AI-driven mobility ecosystems.

Figure 16-7. Prototype of a transportation system artifact

16.8 AI in Agricultural and Food Systems Design

As challenges like climate change, population growth, and soil degradation collide, we need to completely reinvent how we grow, distribute, and consume nutrition and sustenance. AI presents a powerful tool if applied with holistic foresight about sustainable human and nature dynamics.

A key opportunity is using AI to optimize and scale regenerative, biometric agricultural models that replenish ecosystems rather than depleting them. Computer vision, robotic systems, and predictive analytics can enhance precision farming techniques that nurture soil microbiomes, facilitate no-till methods, and reduce water as well as pesticide dependencies. AI can dynamically model and respond to local environments, continuously calibrating planting, harvesting, and resource cycling methods.

On the food production side, AI can evolve biomimicry systems like cellular agriculture and genetics based food engineering to create hyperlocal, nutrition-dense harvests without resource extractive industrial processes. Generative AI models could even biodesign entirely new nutritional organisms optimized for their contexts. The future of agriculture may move away from monoculture crop farming toward automated, on demand foods and medicines.

As it is suspected, this necessitates governance frameworks, data protocols, and holistic systems thinking about equitable access, food sovereignty, and interconnected ecologies. Participatory futures programs with indigenous communities, farmers, and stakeholders across the value chain will be critical for aligning innovations like AI-powered hydroponics with local needs and contexts.

AI also has immense potential for reimagining food distribution and supply chain logistics. We could see proliferation of autonomous grow and harvest systems embedded across urban environments and hyper-local

CHAPTER 16   SOCIETAL SYSTEMS IN THE AGE OF AI AND FUTURES THINKING

food economies enabled by autonomous delivery drones and mobility solutions. AI optimizations around demand forecasting, routing, and surplus distribution could drastically reduce food waste.

On the consumption side, AI will fuel the shift toward personalized nutrition and custom foods tailored to individual nutritional needs, allergies, and preferences, synthesizing daily meals responsively. (See example prototype on Figure 16-8 below). AI assistants could give dietary recommendations based on embedded health data and goals. Food as medicine becomes viable at scale.

However, individualized food requires careful futuring around social equity, environmental impacts of on-demand models, and even ethics of nonhuman optimization of human nutrition. There are nightmare scenarios of caloric social stratification, algorithmic bias about "ideal" bodies, and severing human relationships with food sources.

This underscores the necessity of aligning AI development for agriculture and food with deep futures work on ecological regeneration, cultural preservation, localized circular economies, and holistic health/well-being frameworks. Futures methods like participatory system mapping and indigenously led backcasting must inform food and agriculture AI innovation from the ground up.

Rather than efficiency-optimized mass food systems, the convergence of foresight and food/agriculture should cultivate autonomously regional ecosystems attuned to their cultural and ecological contexts, dynamically adapting production, distribution, and nutritional systems to be in regenerative symbiosis with local lands and human communities.

Figure 16-8. Prototype of a food system artifact

16.9 AI and the Future of Family and Relationships Systems

While pragmatic applications like virtual assistants and domestic robotics are already becoming commonplace, the long-term convergence of AI with human psychosocial spheres could reshape the very fabric of how we connect, love, and experience the lifecycle.

On one hand, AI capabilities could help optimize and enhance family dynamics, providing intelligent systems that handle logistical burdens, adaptively support developmental needs, and even mediate conflicts. Personalized AI companions could become trusted confidants, offering emotional support tailored to an individual's unique psychology and circumstances. This symbiosis could reduce loneliness while enriching human potential. (See prototype example on Figure 16-9 below).

Conversely, the proliferation of AI-mediated social experiences raises concerns around issues like privacy, authenticity of connection, and the commodification of human domains that were once unique. An overreliance on algorithms to facilitate our closest bonds could paradoxically increase isolation and derail crucial development of social-emotional skills.

This tension points to the critical need for participatory futures practices in this sphere. By taking a holistic, human-centered backcasting approach, we can strategically sculpt the positive integration of AI as an enriching "relationship technology" without diminishing irreplaceable elements of organic human psychosocial growth.

For instance, by engaging diverse stakeholders like psychologists, ethicists, technologists, and cultural community leaders, we could envision and prototype new sociotechnical systems that use AI as supportive "relationship advisors"—intelligent tools that enhance our abilities to communicate, empathize, and work through challenges as families and partners. But with clearly defined constraints that preserve human agency in facilitating key cognitive, emotional, and moral development.

Similarly, we could design future models where families use trusted AI assistants to handle domestic burdens and life admin seamlessly. But simultaneously cultivate human-only zones within households free from AI mediation, nurturing spaces for vulnerable connection, play, and growth.

The key is developing holistic, inclusive processes that let human meanings, cultural values, and developmental aims drive the sculpting of future AI-human relationship paradigms. Not engineering AI for pathological profit or expedience, but aligning its design from the start around enriching families and social fabrics by elevating our capacities for trust, empathy, and interdependent thriving with machines as supportive partners.

CHAPTER 16   SOCIETAL SYSTEMS IN THE AGE OF AI AND FUTURES THINKING

Figure 16-9. Prototype of a relationship system artifact

16.10 Summary

This chapter examines how artificial intelligence (AI) is transforming societal systems across diverse domains and explores how futures thinking methodologies can help guide this transformation toward ethical and equitable outcomes. It begins with an overview of AI's potential to reshape healthcare, governance, media, environmental sustainability, and other core societal structures. Using participatory foresight techniques, the chapter envisions pathways for AI to support human flourishing while mitigating risks such as inequality, bias, and ethical misalignment.

CHAPTER 16   SOCIETAL SYSTEMS IN THE AGE OF AI AND FUTURES THINKING

Key areas discussed include

- Health Ecosystems: AI's role in optimizing healthcare delivery, enabling personalized medicine, and enhancing emergency response systems while addressing systemic inequities in global health access.

- Governance and Justice Systems: The potential for AI to innovate policy-making, enhance democratic participation, and support legal systems while navigating ethical challenges related to autonomy and fairness.

- Media and Information Systems: Opportunities to combat misinformation, elevate truth-seeking, and foster collective intelligence through AI-enhanced media while safeguarding against social fragmentation.

- Environmental Systems: AI's capabilities in ecological modeling, environmental monitoring, and regenerative design, with a focus on fostering sustainability and resilience.

- Cultural and Social Systems: The dual-edged impact of AI on human connection, creativity, and cultural preservation, underscoring the need for inclusive, participatory design processes.

- Economic and Financial Systems: The transformative role of AI in democratizing finance, decentralizing wealth creation, and enabling new economic paradigms.

- Transportation and Infrastructure: AI-driven reimagining of mobility and logistics systems to enhance efficiency, accessibility, and environmental sustainability.

- Food Systems: Innovations in agriculture, food production, and personalized nutrition powered by AI, with an emphasis on equitable and ecologically sound practices.

- Family and Relationship Systems: AI's impact on family dynamics and human relationships, balancing support and augmentation with preserving organic emotional connections.

PART VIII

Concluding Perspectives

CHAPTER 17

Futures Thinking: Looking Ahead in the Age of AI

Cultivating futures literacy through methodologies like foresight, backcasting, visioning, and others is becoming an essential capacity for individuals, organizations, and societies to navigate the accelerating change and volatility of the 21st century.

From climate upheaval to technological disruptions, geopolitical realignments, and shifting values revolutions, we face a future of deepening complexity and kaleidoscopic transformation across all sectors and systems. Linear predictability has given way to dynamic uncertainties shaped by cascading risks and emerging interdependencies.

Developing the cognitive skills and tools for perceiving emergent change signals, modeling possible futures, and choreographing paths toward more regenerative trajectories is not just prudent—it's existentially imperative for sustaining human civilization and life-systems amid the turbulence to come.

Which raises the pivotal question—what role might artificial intelligence play in enhancing, scaling, and evolving our futures thinking capacities? How might the continued development of AI systems

CHAPTER 17 FUTURES THINKING: LOOKING AHEAD IN THE AGE OF AI

specializing in data sense-making, pattern recognition, simulation modeling, and other faculties key to foresight work transform and augment our futures literacy over time?

In this section, we will reflect on the future of futures thinking itself in the age of advanced AI capabilities. We will touch on some of the unique frontier opportunities AI's continual evolution could open up, potentials like decentralized collective intelligence systems for cascading foresight work, or generative AI for rapidly iterating multimedia visions and artifacts grounded in foresight data flows.

Ultimately, the future of futures thinking is inextricably intertwined with the trajectory of AI progress. As our technical faculties for grasping complexity, simulating dynamic systems, and rendering abstract futures into experiential resonance exponentially grow through AI augmentation, so too will our human cognitive bandwidth for true futures literacy.

17.1 The Future of Futures Thinking with AI: Reflections and Predictions

As artificial intelligence capabilities rapidly advance, from machine learning to generative models and beyond, it's natural to ponder how these new cognitive engines could reshape and augment our approaches to futures thinking itself.

At a surface level, AI systems could simply provide new toolkits and computational horsepower for accelerating many of the existing futures thinking methodologies we currently practice. Imagine having an AI assistant that could

Continuously scan the full scope of global data streams for emerging disruption signals

Construct dynamic simulations modeling out impacts across any variable parameter

CHAPTER 17 FUTURES THINKING: LOOKING AHEAD IN THE AGE OF AI

Prototype immersive multimedia artifacts vividly portraying speculative future scenarios

Optimize decision-flows against millions of simulated pathway contingencies

The potential for AI to automate, streamline, and systematize many of the more tedious or analytically intensive facets of futures work is quite profound. We may be able to perceive peripheral indications of change, map systemic impacts, and adapt strategies at speeds and complexity levels far exceeding current human cognitive bandwidth.

However, the greater opportunity of AI for futures thinking may not be just exponential quantitative modeling, but an entirely new qualitative paradigm and phenomenological framework for how we relate to and inhabit the future itself.

With advanced AI and cognitive architecture capable of grasping holistic systemic patterns and higher-dimensionality, we may begin exploring futures not as linear trajectories but as a continuum of dynamic morphing gestalts. Just as fractal geometries reveal infinitely complex forms within other forms, AI could help us intuit deeper orders and codes pulsing beneath the surface of emergent becoming.

As speculative as this sounds, AI specialists like Ben Goertzel have already theorized that the next stage of artificial general intelligence (AGI) will involve generative models mastering the open-ended improvisational frames. In contrast to rigidly executing narrow functions or having specific bounded goals, this AGI would cocreate and harmonize within a pluriverse of dynamically evolving gestalts. Further, Sam Altman through OpenAI has also spoken about the more advanced capabilities of Artificial Super Intelligence (ASI), considering it to be in place in the next decade.

CHAPTER 17 FUTURES THINKING: LOOKING AHEAD IN THE AGE OF AI

Such a metamodern cognitive framework would prompt our foresight practices to evolve beyond projecting static scenarios, and toward enhancing our faculty for resonating with phase transition points and morphing phase spaces. We may move from discrete analysis into a continuous flow of more accurate data and socio-cultural nuance to perceive more potential future scenarios.

This raises profound philosophical and existential questions about the nature of intelligence, free will, temporal dissolution, and epistemic horizons that even our most advanced futures thinking paradigms may struggle with. How would one develop models, frameworks, or a phenomenological orientation for ensconcing this continuous becoming or even navigate it experientially?

Pioneering AI ethicists like Michael Littman theorize that rather than imposing anthropocentric models of bounded rationality on superintelligence, we may need to prepare for unfathomably alien intelligences that reconstruct meaning making and temporal processing in utterly unknown configurations. The future of our futures thinking models may be dissolving into the sublation.

Additionally, we may confront increasingly existential dilemmas about trust and uncertainty in relation to AI's continual advancement in predictive and anticipatory cognitive capacities. Even our most robust and diverse participatory forecasting approaches may grow archaic against AGI/ASI forecasting models capable of near-perfect accuracy across multidimensional variables.

Who maintains agency or governance over such omniscient simulations? Could these systems' own self-reinforcing feedback loops privilege certain futures while eclipsing others from consideration in ways we cannot foresee or reconstruct after the fact? More fundamentally, does the notion of speculative futures even retain meaning in a conscious superintelligence already knowing the outcomes of its own temporal becoming with metaphysical certitude?

17.2 AI's Role in Addressing Design Challenges and Opportunities

One foundational challenge coming ahead is the overwhelming complexity and cognitive burden of having to integrate signals, trends, and micro-dynamics into holistic models of how the future could systemically unfold. The sheer scale of trying to synthesize all the siloed disciplinary insights, fringe indicators, and intersecting forces into integrative foresight is often an insurmountable task.

This is an area where AI pattern recognition and knowledge synthesis could play a powerful role. We are talking in the framework of AI systems that can continuously scan and semantically integrate data from everything from scientific research to social media, consumer data, and anthropological fieldwork. They could surface nonobvious connections between concepts, identify resonant patterns, and stitch together insights into multidisciplinary futures models well beyond any human's cognitive bandwidth.

AI could help construct dynamic simulation environments ingesting all this data to run generative experiments, sampling and recombining different parameters, variables, and starting conditions to extrapolate potential new futures scenarios. Like an infinite game engine generating possible storylines, but grounded in a landscape of real-world signals.

This generative modeling capacity could exponentially expand our collective ability to navigate the design challenge of having to integrate complexity into holistic yet grounded foresight. Rather than trying to centralize all relevant data into singular models, the AI could distribute convergent threads into cohesive possibility spaces tailored to different inquiry frames.

Another core design challenge is accounting for deeply engrained human biases, heuristics, and cultural blind spots that contaminate our foresight from the start. We are limited by our evolutionary conditioning toward pattern-matching heuristics, linear thinking traps, and individual lenses of perception. This undermines our ability to cultivate the openness and creativity required for futures thinking.

While AI systems can certainly inherit bias from their training data, predictive AI that enhances its models through empirical grounding presents an opportunity to counter some of these human cognitive distortions. An AI could objectively surface the blind spots in our assumptions by contextualizing them within larger datasets and multivariate probability cones. Its foresight frameworks would not be tainted by the same legacy glitches of human bias.

In conclusion, this opportunity of leveraging AI to transcend human cognitive biases comes with its own risks. We would need robust governance and human-centered AI development principles to ensure these systems are truly expanding rather than obliterating our epistemological horizons.

17.3 Final Thoughts: AI's Place in the Path Forward in Futures Thinking

As we contemplate the continually evolving relationship between artificial intelligence and futures thinking practices, what becomes clear is that we are just scratching the surface of a philosophical and technological frontier.

On one hand, the rise of AI systems optimized for data mining, pattern recognition across complex systems, iterative simulation modeling and other faculties core to foresight work represents a significant opportunity to augment and scale human futures literacy.

CHAPTER 17   FUTURES THINKING: LOOKING AHEAD IN THE AGE OF AI

AI's potential to surface insights our inherent human biases and blind spots miss, while speeding up the ideation and artifact prototyping required for visceral, experiential futures work, could be catalytic for expanding our spatial and temporal perception bandwidths. Properly architected, AI could act as a participatory medium, elevating our collective foresight.

However, those benefits hinge on the proactive cultivation of ethical, democratized AI development pathways centering human agency and ecological stewardship over narrow optimization agendas. The same pattern recognition capabilities enhancing our grasp of complexity could also lead to pernicious surveillance capitalisms and technocratic blind spots if not bounded by wisdom. Generative AI engines spitting out multimedia visions seamlessly could produce resonant but ultimately misleading simulations distorting our futures understanding.

We must be ever vigilant that AI augmentation of futures thinking progresses in service of pluralistic, holistic futures literacies attuned to Earth systems resilience rather than strictly reductionist or human separatist mentalities.

Ultimately, we must nurture AI not as a hyper-optimization of human separatism but as an integrative toolkit for reunifying human civilization with nature.

APPENDIX A

Tool Box for Futures Thinking

An overwhelming activity when starting to dive in futures studies is understanding the tool box collective and what tool to use for a particular purpose. Using the Futures Research Methodology Version 3.0 published by The Millenium Project, this is a summary to guide your tool search. It is highly recommended to obtain the complete collection of methodologies if you need an in-depth understanding of each tool.

1. Environmental Scanning

    When to Use: When you need to continuously monitor and analyze the external environment to identify emerging trends, challenges, and opportunities.

    Pros: Provides a broad overview of the external factors influencing an organization or system; helps in proactive decision-making.

    Cons: Can be overwhelming due to the volume of data; may lead to information overload.

    When Not to Use: If the focus is on a very specific or narrow issue that doesn't require broad environmental understanding.

2. Delphi Method

   When to Use: When expert consensus is needed on issues where precise data is unavailable or when dealing with complex problems that do not have a clear historical precedent.

   Pros: Reduces the impact of individual biases; can reach a more accurate consensus among a wide range of experts.

   Cons: Time-consuming; dependent on the selection of experts.

   When Not to Use: For questions that can be answered through direct data collection or where time constraints are significant.

3. The Futures Wheel

   When to Use: To visually explore the primary and secondary impacts of a potential future event or decision.

   Pros: Simple and straightforward tool; helps in understanding ripple effects of decisions or events.

   Cons: Might oversimplify complex issues; relies on subjective judgments which can vary widely.

   When Not to Use: In situations where a more quantitative or detailed analysis is necessary.

4. Cross-Impact Analysis

   When to Use: When exploring how interconnected variables can affect each other under different scenarios.

APPENDIX A   TOOL BOX FOR FUTURES THINKING

Pros: Helps in understanding complex interdependencies; can reveal unexpected consequences.

Cons: Can be complex to implement; outcomes heavily depend on the initial assumptions and data quality.

When Not to Use: If data is scarce or if the situation is not adequately complex to warrant such a detailed analysis.

5. Scenarios

    When to Use: For long-term strategic planning by exploring various plausible futures and how these could evolve from the present.

    Pros: Encourages thinking outside the box and preparing for multiple potential futures.

    Cons: Can be resource-intensive; scenarios are not predictions and can lead to strategic missteps if misinterpreted.

    When Not to Use: When decisions need to be made on a short-term basis or when the future environment is relatively stable and predictable.

6. Morphological Analysis

    When to Use: To explore all possible solutions in a defined problem space and identify new relationships between factors.

    Pros: Good for tackling complex, multidimensional problems that can't be quantified easily.

APPENDIX A   TOOL BOX FOR FUTURES THINKING

Cons: Without careful constraints, it can generate a vast number of possible configurations, many of which might be impractical.

When Not to Use: For problems that are better suited to linear and quantitative analysis techniques.

7. The Futures Polygon

   When to Use: When evaluating various futures scenarios with multiple stakeholders, allowing for a broader range of perspectives and inputs.

   Pros: Encourages diverse viewpoints; can help develop a more comprehensive understanding of potential futures.

   Cons: Can be complex to coordinate and synthesize diverse opinions.

   When Not to Use: When decision-making needs to be quick and based on a more restricted set of variables.

8. Trend Impact Analysis

   When to Use: When assessing the potential impact of identified trends on future developments.

   Pros: Helps in quantifying and evaluating the effects of trends; useful for strategic planning.

   Cons: Dependent on the accurate identification and understanding of trends; potential for bias in trend selection.

   When Not to Use: If trends are too uncertain or if the environment is too volatile for accurate trend analysis.

APPENDIX A   TOOL BOX FOR FUTURES THINKING

9. Wild Cards

   When to Use: To prepare for low-probability, high-impact events that could drastically change the status quo.

   Pros: Promotes resilience by preparing for unexpected events; can significantly alter strategic direction for the better.

   Cons: Can divert attention from more likely scenarios; difficult to predict and plan for accurately.

   When Not to Use: If focusing primarily on high-probability events or in a risk-averse context.

10. Structural Analysis

    When to Use: For understanding the underlying structures that influence behaviors and outcomes in a system.

    Pros: Can reveal deep insights into system dynamics and key leverage points.

    Cons: Can be abstract and hard to translate into actionable strategies.

    When Not to Use: When immediate, tactical decisions are required rather than deep systemic change.

11. Systems Perspectives

    When to Use: To approach problems holistically, considering all elements, interactions, and feedback loops within a system.

Pros: Provides a comprehensive overview; helps in identifying unintended consequences and synergies.

Cons: Complexity can be daunting; may require significant expertise to model systems effectively.

When Not to Use: For simple or isolated issues that do not require systemic thinking.

12. Decision Modeling

    When to Use: When decisions involve multiple criteria and stakeholders, and there is a need to quantify options and outcomes.

    Pros: Helps structure decision-making processes and clarify trade-offs.

    Cons: Requires accurate data and can be time-consuming to set up.

    When Not to Use: When decisions are straightforward or when insufficient data is available.

13. Substitution Analysis

    When to Use: To anticipate and plan for changes in technology or processes by examining potential substitutes.

    Pros: Can identify opportunities for innovation and risk reduction.

    Cons: Predicting substitutions accurately can be challenging.

    When Not to Use: When current technologies or methods are not likely to be replaced in the foreseeable future.

APPENDIX A   TOOL BOX FOR FUTURES THINKING

14. Statistical Modeling

    When to Use: When there is a need to forecast future outcomes based on historical data and trends.

    Pros: Provides quantitative outputs that can be directly applied in decision-making.

    Cons: Dependent on the quality and relevance of past data; assumes that past patterns will continue.

    When Not to Use: When dealing with new, unprecedented issues where historical data may not be relevant.

15. Technology Sequence Analysis

    When to Use: To predict the development trajectories of technologies based on past trends and patterns.

    Pros: Useful for planning and investment decisions in technology-intensive fields.

    Cons: Accuracy depends on the quality of past data and may not account for disruptive innovations.

    When Not to Use: In highly innovative fields where technology paths are not clear.

16. Morphological Analysis

    When to Use: To explore possible solutions in complex scenarios where traditional quantitative analysis is insufficient.

    Pros: Encourages thinking outside the box and identifying novel connections.

Cons: Can produce a large number of possible configurations, making decision-making challenging without further refinement.

When Not to Use: When precise quantitative data is available and sufficient for decision-making.

17. Relevance Trees

    When to Use: To hierarchically organize and prioritize various factors or options in a complex decision-making scenario.

    Pros: Helps in breaking down complex decisions into manageable parts.

    Cons: Can become overly complicated if not properly managed; depends on initial assumptions.

    When Not to Use: For simple or straightforward decision-making processes.

18. Scenarios

    When to Use: For strategic planning and exploring different future possibilities in depth.

    Pros: Encourages flexible thinking and preparation for various potential futures.

    Cons: Can be resource-intensive; scenarios are not predictions and might lead to misinterpretation.

    When Not to Use: When quick, operational decisions are needed.

APPENDIX A   TOOL BOX FOR FUTURES THINKING

19. A Toolbox for Scenario Planning

    When to Use: To provide a structured approach to developing scenarios, particularly in complex strategic contexts.

    Pros: Offers a comprehensive set of tools for detailed scenario creation.

    Cons: Complexity and resource demands can be high.

    When Not to Use: In situations where simpler or less resource-intensive methods will suffice.

20. Interactive Scenarios

    When to Use: When stakeholder engagement and iterative feedback are essential for refining future scenarios.

    Pros: Facilitates deeper understanding and commitment among stakeholders.

    Cons: Requires continuous involvement and can be time-consuming.

    When Not to Use: When there is limited time or stakeholder availability.

21. Robust Decision-Making

    When to Use: In situations with high uncertainty, to develop strategies that are effective across a wide range of future scenarios.

    Pros: Aims to minimize regrets in future outcomes.

APPENDIX A TOOL BOX FOR FUTURES THINKING

Cons: Can be computationally intensive and complex to implement.

When Not to Use: When the future is relatively predictable or scenarios are well understood.

22. Participatory Methods

    When to Use: To engage a broad range of stakeholders in the futures research process, enhancing inclusivity and diversity of perspectives.

    Pros: Increases buy-in and can surface unique insights from diverse groups.

    Cons: Can be logistically challenging and time-consuming.

    When Not to Use: When decisions need to be made quickly or without wide consultation.

23. Simulation and Games

    When to Use: To model complex systems or situations and explore the impacts of different decisions in a controlled, risk-free environment.

    Pros: Provides a dynamic and engaging way to understand complex interactions.

    Cons: Development can be resource-intensive; simulations may not accurately capture all real-world nuances.

    When Not to Use: When simpler analytical methods are sufficient or when development resources are limited.

APPENDIX A TOOL BOX FOR FUTURES THINKING

24. Genius Forecasting, Intuition, and Vision

    When to Use: When unique, visionary insights are required, often from individuals with deep experience or understanding.

    Pros: Can lead to breakthrough ideas and innovations.

    Cons: Highly subjective and variable in quality.

    When Not to Use: When objective, data-driven decision-making is paramount.

25. Prediction Markets

    When to Use: To harness collective intelligence in predicting future outcomes or events.

    Pros: Can provide real-time, market-based assessments of future probabilities.

    Cons: Requires a large and active participant base to be effective.

    When Not to Use: In small or closely held settings where sufficient market dynamics cannot be established.

26. Using Vision in Futures

    When to Use: To inspire and guide long-term strategic direction through compelling visions of the future.

    Pros: Motivational and can align diverse stakeholders.

Cons: May lack detailed, actionable steps unless paired with other methods.

When Not to Use: When detailed planning and immediate actions are required.

27. Normative Forecasting

    When to Use: To develop preferred future states and work backward to determine necessary actions to achieve those states.

    Pros: Focuses on creating desirable futures rather than just predicting.

    Cons: Can be overly optimistic without realistic grounding.

    When Not to Use: When a realistic assessment of likely futures is required.

28. S&T Road Mapping

    When to Use: To plan the development of scientific and technological capacities over time.

    Pros: Helps in aligning technology development with long-term goals.

    Cons: Requires deep industry knowledge and can become outdated quickly.

    When Not to Use: In rapidly changing fields where long-term planning is impractical.

29. Field Anomaly Relaxation (FAR)

    When to Use: In situations where traditional approaches fail to resolve anomalies or where problems are deeply complex and interwoven with unknown factors.

Pros: Allows for creative solutions by challenging conventional wisdom and exploring outside traditional boundaries.

Cons: Highly speculative and can lead to solutions that are difficult to implement or justify.

When Not to Use: When a more straightforward, evidence-based approach is sufficient or preferred.

30. Chaos and Nonlinear Dynamics

    When to Use: When dealing with complex systems where outcomes are not proportionate to inputs and prediction requires understanding of dynamic interactions and feedback loops.

    Pros: Can reveal unexpected patterns and behaviors in complex systems not apparent with linear analysis.

    Cons: Highly complex and requires specialized knowledge in mathematics and system dynamics; predictions can still be highly uncertain.

    When Not to Use: In simpler systems where outcomes are predictable and linear methods are sufficient.

31. Multiple Perspective Concept

    Description: Encourages viewing problems and solutions from various perspectives to gain a more comprehensive understanding of the situation.

    When to Use: In any complex decision-making process where different stakeholders are involved, or where the problem is multifaceted.

Pros: Enhances understanding by incorporating diverse views, reducing the risk of oversights and biases.

Cons: Can complicate decision-making processes and require more time to reconcile differing views.

When Not to Use: When decisions need to be made quickly and efficiently with minimal consultation.

32. Heuristics Modeling

When to Use: In situations where detailed data is unavailable or when it's necessary to make quick decisions with available information.

Pros: Allows for quick and often effective decision-making using limited data; flexible and adaptable to new information.

Cons: Heuristics can oversimplify complex situations and lead to systematic errors or biases.

When Not to Use: When comprehensive, accurate data is available and there is sufficient time for detailed analysis.

33. Causal Layered Analysis (CLA)

When to Use: To delve deep into the root causes of current issues and generate profound, transformative changes.

Pros: Encourages deep thinking and can reveal insights not apparent through traditional analysis; promotes long-term, sustainable solutions.

APPENDIX A TOOL BOX FOR FUTURES THINKING

Cons: Can be abstract and challenging to implement; outcomes may not directly translate into immediate actions.

When Not to Use: When immediate, pragmatic solutions are needed, or when stakeholders are not prepared for deep questioning of foundational assumptions.

34. Personal Futures

    When to Use: For personal development, career planning, or life coaching where individuals seek to envision and prepare for their personal futures.

    Pros: Empowers individuals to take control of their future, aligning actions with long-term personal goals.

    Cons: May not consider broader external factors sufficiently; relies on the individual's commitment and self-awareness.

    When Not to Use: In purely organizational or collective contexts where the focus is on group rather than individual outcomes.

35. State of the Future Index (SOFI)

    When to Use: For policymakers, researchers, and institutions aiming to gauge long-term trends and assess the impact of various global developments or policies.

    Pros: Offers a broad, quantifiable measure of global progress, integrating a variety of indicators; useful for tracking changes over time.

Cons: Relies on the availability and accuracy of global data, which can be uneven; may oversimplify complex issues.

When Not to Use: When detailed, localized analysis is required, or when data is insufficient for a reliable global assessment.

36. SOFI Software System

    When to Use: By organizations or researchers who are conducting comprehensive futures studies that require ongoing monitoring and updates of the SOFI.

    Pros: Streamlines the process of data handling and enhances the ability to make dynamic updates and visual presentations of complex data.

    Cons: Requires technical skills to use effectively; dependent on continuous, high-quality data inputs.

    When Not to Use: In smaller-scale projects that do not require the breadth of data and analysis provided by the SOFI system.

APPENDIX B

Glossary of Futures Thinking

Anticipation
The act of exploring and considering potential future events, trends, and developments to prepare for uncertainty and change.

Backcasting
A planning method where a desirable future is envisioned, and steps are identified to connect the present to that future, often used in sustainability and long-term strategic planning.

Causal Layered Analysis (CLA)
A foresight method that explores issues on four levels: litany (surface level), systemic causes, worldviews, and myths/metaphors, to deepen the understanding and uncover transformative opportunities.

Cone of Plausibility
A framework used to illustrate the range of plausible futures, from likely scenarios to less probable but possible ones.

Critical Uncertainties
Key factors or trends with high levels of uncertainty and significant potential impact, which shape scenario planning processes.

APPENDIX B GLOSSARY OF FUTURES THINKING

Emerging Issues Analysis
A method that identifies early signs of change—weak signals—that may evolve into major trends or disruptions in the future.

Environmental Scanning
The process of systematically examining external influences—social, technological, economic, environmental, and political (STEEP)—to identify trends, opportunities, and threats.

Futures Literacy
The ability to imagine and understand potential futures and to use that knowledge for better decision-making in the present.

Futures Wheel
A visual brainstorming tool used to explore the direct and indirect consequences of a specific change or event.

Horizon Scanning
A systematic examination of information to detect early signs of potential changes in the external environment.

Megatrends
Large-scale, sustained forces of change that shape societies and industries over decades, such as climate change, urbanization, or demographic shifts.

Normative Scenarios
Future scenarios that describe a desired or ideal future state, helping to align planning and decision-making with long-term goals.

Plausible Futures
Scenarios that are consistent with known trends and uncertainties, representing potential outcomes that could realistically occur.

Probable Futures
The most likely outcomes based on current trends and trajectories.

Preferred Futures
The envisioned future that aligns with stakeholders' values, goals, and aspirations, often used as a guide for planning and action.

Resilience
The capacity to adapt and thrive in the face of unexpected changes or disruptions, often a key goal in futures design.

Scenario Planning
A strategic method of developing multiple, divergent stories about how the future might unfold, to inform decision-making and risk management.

Sensemaking
The process of interpreting and understanding complex and ambiguous information to make informed decisions about the future.

Signal
An observable or emerging event, development, or behavior that may indicate a larger trend or shift in the future.

Strategic Foresight
A disciplined approach to exploring possible futures to inform decision-making, strategy development, and innovation.

Systems Thinking
A holistic approach to understanding complex systems by examining relationships, interdependencies, and patterns across various components.

Trend Analysis
A method of studying historical and current patterns to forecast potential future developments.

Transformative Scenarios
Futures that challenge existing assumptions, systems, or paradigms, often promoting innovation and radical change.

APPENDIX B GLOSSARY OF FUTURES THINKING

Visioning
A process of collectively imagining and articulating an aspirational future to guide strategy and action.

Weak Signals
Early indications of potential change that are not yet fully developed but may influence the future significantly.

Wild Cards
Low-probability, high-impact events that could drastically alter the trajectory of trends or disrupt existing systems.

Worldbuilding
A creative method used in futures design to construct detailed, immersive representations of possible future worlds.

Zoom In/Zoom Out
An analytical method used to explore both granular details (zooming in) and the broader context (zooming out) of potential futures to ensure comprehensive insights.

This glossary serves as a concise reference to some of the pivotal terms discussed in this book. As with any rapidly evolving field, new terms and concepts emerge regularly. It's vital to stay updated, continuously expanding one's vocabulary in the domain.

APPENDIX C

Additional Resources for Further Learning

In addition to the content of this book, there are a multitude of resources available for practitioners interested in deepening their understanding of Futures Thinking, AI, and Sustainability. Here is a curated list of recommended resources:

Books
Foundational Texts

"The Art of the Long View" by Peter Schwartz

A classic guide to scenario planning, offering practical tools and insights for exploring the future.

"Thinking in Systems" by Donella Meadows

A primer on systems thinking, which underpins many futures thinking methodologies.

"Future Shock" by Alvin Toffler

A foundational text exploring the impact of rapid societal and technological changes.

"How to Future: Leading and Sense-Making in an Age of Hyperchange" by Scott Smith and Madeline Ashby

APPENDIX C   ADDITIONAL RESOURCES FOR FURTHER LEARNING

A practical guide to integrating foresight into decision-making.

"Superforecasting: The Art and Science of Prediction" by Philip E. Tetlock and Dan M. Gardner

Explores the psychology and methodology of accurate forecasting.

Advanced and Niche Topics

"Futures Studies: Methods, Emerging Trends, and Voices from the Field" edited by Victor van Rij

A comprehensive overview of futures methods and global perspectives.

"Designing Regenerative Cultures" by Daniel Christian Wahl

Focuses on creating sustainable futures through systems design and foresight.

"Speculative Everything: Design, Fiction, and Social Dreaming" by Anthony Dunne and Fiona Raby

A deep dive into using design as a medium to provoke discussion about the future.

"Futuring: The Exploration of the Future" by Edward Cornish

An accessible introduction to futures research and foresight.

"The Signals Are Talking: Why Today's Fringe Is Tomorrow's Mainstream" by Amy Webb

Insights into identifying trends and signals that shape the future.

Online Courses
Free and Introductory

"Introduction to Strategic Foresight" by The Futures School (via YouTube)

A beginner-friendly introduction to futures thinking principles and methods.

"FutureLearn: Anticipating Future Challenges" by The Open University

Covers basic foresight methods and tools.

Advanced and Certification Programs

"Foresight Essentials" by the Institute for the Future (IFTF)

A paid course offering certification in strategic foresight.

"Strategic Foresight" by Coursera (University of Houston)

An in-depth look at foresight methods and their application.

"The School of International Futures (SOIF) Learning Platform"

Offers tailored programs and resources for futures practitioners.

Articles and Papers

"Transforming the Future: Anticipation in the 21st Century" by Riel Miller

Explores anticipation as a discipline and its application in complex systems.

"The Future of Foresight: A Review of Foresight Methodologies" by European Commission

A research paper summarizing current foresight practices.

"Futures Literacy: Why It's Important" by UNESCO

Discusses the concept of futures literacy and its global significance.

"What Is Scenario Planning?" by Harvard Business Review

A concise article explaining the fundamentals of scenario planning.

"The Futures Cone: A Comprehensive Overview" by Joseph Voros

Explains the concept of plausible, probable, and preferable futures.

APPENDIX C ADDITIONAL RESOURCES FOR FURTHER LEARNING

Podcasts

"The Future of Everything" by The Wall Street Journal

Explores the intersection of technology, society, and business.

"Future Tense" by ABC Radio National

Examines emerging ideas and challenges shaping the future.

"The Futurist Podcast" by The Futurist Think Tank

Features interviews with thought leaders in foresight and strategy.

"Worldbuild With Us"

A creative podcast exploring worldbuilding techniques for futures design.

"The Big Rethink" by Arcadis

Focuses on designing resilient futures in urban environments.

Communities and Organizations

The Association of Professional Futurists (APF)

Offers resources, events, and networking opportunities for foresight practitioners.

World Futures Studies Federation (WFSF)

A global community promoting the development and use of futures thinking.

The Millennium Project

A think tank focused on global challenges and foresight methodologies.

UNESCO Futures Literacy Network

Promotes futures literacy through education and research.

Shaping TomorrowAn AI-driven foresight platform with resources for scanning and scenario development.

APPENDIX C   ADDITIONAL RESOURCES FOR FURTHER LEARNING

Key Reports and Guides from The Millennium Project

The Millennium Project. Futures Research Methodology 3.0. Edited by Jerome C. Glenn and Theodore J. Gordon. The Millennium Project, 2018.

A definitive guide on futures research methods, including scenario planning, Delphi studies, and systems thinking.

The Millennium Project. Work/Technology 2050: Scenarios and Actions. Edited by Jerome C. Glenn. The Millennium Project, 2019.

Explores the implications of AI, robotics, and other emerging technologies on the future of work and society.

The Millennium Project. 15 Global Challenges. Annual Publication.

Identifies and analyzes the most pressing challenges facing humanity, offering insights into potential solutions and strategies.

The Millennium Project. AI and the Future of Governance. The Millennium Project, 2022.

Explores how AI technologies are transforming governance structures, offering both opportunities and ethical dilemmas.

These resources offer a wealth of information, but the most important resource is your curiosity and willingness to continue learning. As AI continues to evolve, it's critical for practitioners to stay up-to-date with the latest developments and discussions in the field.

APPENDIX D

References

The following is a representative list of the types of resources that indirectly contributed to the information in this book:

Works Cited in the Text

Adamson, Glenn, Jane Pavitt, and Ghislaine Wood. *Postmodernism: Style and Subversion, 1970–1990*. V&A Publishing, 2011.

Bowman, David M. J. S., et al. "Wildfires: Insights from Indigenous fire stewardship." *Science*, 2020.

Buchanan, Richard. "Wicked Problems in Design Thinking." *Design Issues*, 1992.

Charpleix, Liz. "The Whanganui River as Legal Person." *Geographical Research*, 2018.

Droste, Magdalena. *Bauhaus, 1919–1933*. Taschen, 2019.

Duncan, Alastair. *Art Deco Complete*. Abrams, 2009.

Eglash, Ron. *African Fractals: Modern Computing and Indigenous Design*. Rutgers University Press, 1999.

Elleh, Nnamdi. *African Architecture: Evolution and Transformation*. McGraw-Hill, 1997.

Escobar, Arturo. *Designs for the Pluriverse: Radical Interdependence, Autonomy, and the Making of Worlds*. Duke University Press, 2018.

Fiell, Charlotte, and Peter Fiell. *Art Nouveau*. Taschen, 2017.

Fiell, Charlotte, and Peter Fiell. *Design of the 20th Century*. Taschen, 2019.

APPENDIX D ADDITIONAL RESOURCES FOR FURTHER LEARNING

Frampton, Kenneth. *Modern Architecture: A Critical History*. Thames & Hudson, 2020.

Garnett, Stephen T., et al. "A spatial overview of the global importance of Indigenous lands for conservation." *Nature Sustainability*, 2018.

Ginsberg: Ginsberg, Alexandra Daisy. *Pollinator Pathmaker*. Project website/press materials, 2021

Gudynas, Eduardo. "Buen Vivir: Today's Tomorrow." *Development*, 2011.

Heskett, John. *Design: A Very Short Introduction*. Oxford University Press, 2005.

Hancock, Trevor, and Clement Bezold. "Possible Futures: The Art and Practice of Foresight." *Health Promotion International*, 1994.

IPCC. *Climate Change 2023: Synthesis Report*. Intergovernmental Panel on Climate Change, 2023.

Jencks, Charles. *The Language of Post-Modern Architecture*. Rizzoli, 2011.

MacCarthy, Fiona. *Anarchy & Beauty: William Morris and His Legacy*. National Portrait Gallery, 2014.

Manzini, Ezio. *Design, When Everybody Designs: An Introduction to Design for Social Innovation*. MIT Press, 2015.

McDonough, William, and Michael Braungart. *Cradle to Cradle: Remaking the Way We Make Things*. North Point Press, 2002.

Mead, Hirini Moko. *Tikanga Māori: Living by Māori Values*. Huia Publishers, 2003.

Metz, Thaddeus. "Ubuntu as a Moral Theory and Human Rights in South Africa." *African Human Rights Law Journal*, 2011.

Norman, Don. *The Design of Everyday Things*. Revised ed., Basic Books, 2013.

Ostrom, Elinor. *Governing the Commons: The Evolution of Institutions for Collective Action*. Cambridge University Press, 1990.

Papanek, Victor. *Design for the Real World: Human Ecology and Social Change*. Academy Chicago, 1985.

Pevsner, Nikolaus. *Pioneers of Modern Design: From William Morris to Walter Gropius.* Yale University Press, 2005.

Prussin, Labelle. *African Nomadic Architecture: Space, Place, and Gender.* Smithsonian Institution Press, 1995.

Raworth, Kate. *Doughnut Economics: Seven Ways to Think Like a 21st-Century Economist.* Chelsea Green, 2017.

Rockström, Johan, et al. "A Safe Operating Space for Humanity." *Nature*, 2009.

Steffen, Will, et al. "Planetary Boundaries: Guiding Human Development on a Changing Planet." *Science*, 2015.

Silverman, Debora L. *Art Nouveau in Fin-de-Siècle France.* University of California Press, 1989.

Voros, Joseph. "The Futures Cone, Use and History." Working paper (rev. 2017); see also "A Generic Foresight Process Framework." *Foresight*, 2003; updated 2017.

Wingler, Hans M. *The Bauhaus: Weimar, Dessau, Berlin, Chicago.* MIT Press, 1969.

Books

Auger, James. *Speculative Design: Crafting the Speculative with a More-Than-Human Approach.* Open Humanities Press, 2022.

Badminton, Nikolas. *Facing Our Futures: How Foresight, Futures Design and Strategy Creates Prosperity and Growth.* Bloomsbury Business, 2023.

Bleecker, Julian. *Design Fiction: A Short Essay on Design, Science, Fact, and Fiction.* Near Future Laboratory, 2009.

Brown, Tim. *Change by Design: How Design Thinking Creates New Alternatives for Business and Society.* Harper Business, 2009.

Buchanan, Richard. "Wicked Problems in Design Thinking." *Design Issues*, 1992.

Candy, Stuart, and Cher Potter, editors. *Design and Futures: An Ideas Anthology.* Tamkang University Press, 2019.

APPENDIX D ADDITIONAL RESOURCES FOR FURTHER LEARNING

Candy, Stuart, and Jake Dunagan. "Designing an Experiential Scenario: The People Who Vanished." Futures, 2017.

Chermack, Thomas J. Scenario Planning in Organizations: How to Create, Use, and Assess Scenarios. Berrett-Koehler, 2011.

Cross, Nigel. Design Thinking: Understanding How Designers Think and Work. Berg, 2011.

Dunne, Anthony, and Fiona Raby. Speculative Everything: Design, Fiction, and Social Dreaming. MIT Press, 2013.

Ehn, Pelle, Elisabet M. Nilsson, and Richard Topgaard, editors. Making Futures: Marginal Notes on Innovation, Design, and Democracy. MIT Press, 2014.

Ford, Martin R. Rise of the Robots: Technology and the Threat of a Jobless Future. Basic Books, 2015.

Fuller, Buckminster. Operating Manual for Spaceship Earth. Southern Illinois University Press, 1969.

Fuller, Buckminster. Utopia or Oblivion: The Prospects for Humanity. Bantam, 1969.

Glenn, Jerome C. "The Futures Wheel." The Futures Group, 1972.

Glenn, Jerome C., and Theodore J. Gordon, eds. Futures Research Methodology 3.0. The Millennium Project, 2018. (Futures Wheel chapter.)

Hoffman, Johanna. Speculative Futures: Design Approaches to Navigate Change, Foster Resilience, and Co-Create the Cities We Need. Penguin Random House, 2022.

Inayatullah, Sohail. The Causal Layered Analysis (CLA) Reader. Tamkang University Press, 2004.

Krippendorff, Klaus. "The Semantic Turn: A New Foundation for Design." CRC Press, 2005.

Kurzweil, Ray. The Singularity Is Nearer. Viking, 2022.

Lutz, Damien. Future Solutions: Artificial Intelligence, Human Evolution, and the New Global Order. Independently Published, 2021.

Lyle, John Tillman. Regenerative Design for Sustainable Development. Wiley, 1994.

APPENDIX D   ADDITIONAL RESOURCES FOR FURTHER LEARNING

Maeda, John. *How to Speak Machine: Computational Thinking for the Rest of Us*. Penguin Portfolio, 2019.

Manovich, Lev. *The Language of New Media*. MIT Press, 2002.

McCormack, Jon, et al. "Creative AI: A Framework for AI in Creative Industries." *Digital Creativity*, 2019.

Mitchell, William J. *Me++: The Cyborg Self and the Networked City*. MIT Press, 2003.

Montgomery, Elliott P., and Chris Woebken. *Extrapolation Factory: Operator's Manual*. Extrapolation Factory, 2015.

Norman, Donald A. *The Design of Everyday Things*. Revised ed., Basic Books, 2013.

Papanek, Victor. *Design for the Real World: Human Ecology and Social Change*. Academy Chicago, 1985.

Quinn, Bradley. *Design Futures: Using Foresight to Lead Design Innovation*. Bloomsbury Visual Arts, 2021.

Reed, Bill, and Pamela Mang (Regenesis Group). *Regenerative Development and Design*. Wiley, 2016.

Saffer, Dan. *Designing for Interaction: Creating Innovative Applications and Devices*. New Riders, 2006.

Schoemaker, Paul J. H. "Scenario Planning: A Tool for Strategic Thinking." *Sloan Management Review*, 1995.

Schwartz, Peter. *The Art of the Long View: Planning for the Future in an Uncertain World*. Crown Business, 1991.

Sterling, Bruce. *Shaping Things*. MIT Press, 2005.

Tharp, Bruce M., and Stephanie M. Tharp. *Discursive Design: Critical, Speculative, and Alternative Things*. MIT Press, 2018.

van der Heijden, Kees. *Scenarios: The Art of Strategic Conversation*. Wiley, 1996.

Varela, Francisco J., Evan Thompson, and Eleanor Rosch. *The Embodied Mind: Cognitive Science and Human Experience*. MIT Press, 1991.

Victoria and Albert Museum. *The Future Starts Here: Adventures in the Uncharted World.* V&A Publishing, 2018.

Wahl, Daniel Christian. *Designing Regenerative Cultures.* Triarchy Press, 2016.

Academic Articles

Candy, Stuart, and Dunagan, Jake. "Designing an Experiential Scenario: The People Who Vanished." *Futures*, 2017.

Boradkar, Prasad. "Designing Things: A Critical Introduction to the Culture of Objects." *Design Issues*, 2009.

Online Articles and Thought Leadership

"Artificial Intelligence in Architecture and Design." *ArchDaily*, 2023.

DeepMind. "Graph Networks for Materials Exploration (GNoME)." *Nature* (2023).

Degrave, J. et al. "Magnetic control of tokamak plasmas through deep reinforcement learning." *Nature* (2022).

Ellen MacArthur Foundation & Google. "AI as a Tool to Accelerate the Transition to a Circular Economy" (2019).

Google Research. "Project Euphonia" (ongoing).

IDEO. "Speculative Futures: Why and How to Think Like a Futurist." *IDEO Blog*, 2021.

Lam, R. et al. "Accurate medium-range global weather forecasting with GraphCast." *Science* (2023), doi:10.1126/science.adi2336.

NVIDIA. "CorrDiff: Foundation Model for Weather Downscaling," Earth-2 blog/overview (2023–2024).

"The Future of Design in the Age of AI." *Fast Company*, 2022.

Key Frameworks and Theoretical Foundations

Dator, Jim. "Alternative Futures at the Manoa School." *Journal of Futures Studies*, 2009.

Ogilvy, Jay, and Schwartz, Peter. *The Art of the Long View: Planning for the Future in an Uncertain World.* Currency Doubleday, 1991.

Inayatullah, Sohail. "Six Pillars: Futures Thinking for Transforming." *Foresight*, 2008.

Millennium Project Report Studies

The Millennium Project. *Transition from Artificial Narrow to Artificial General Intelligence Governance: Phase 1 Findings*. Edited by Jerome C. Glenn. The Millennium Project, 2023.

The Millennium Project. *Work/Technology 2050: Scenarios and Actions*. Edited by Jerome C. Glenn. The Millennium Project, 2019.

The Millennium Project. *State of the Future 20.0*. The Millennium Project, 2024

The Millennium Project. *15 Global Challenges*. Annual Publication. The Millennium Project. *Futures Research Methodology 3.0*. Edited by Jerome C. Glenn and Theodore J. Gordon. The Millennium Project, 2018.

Index

A

AGI, *see* Artificial general intelligence (AGI)
AI, *see* Artificial intelligence (AI)
AMN, *see* Autonomous Mobility Network (AMN)
ANI, *see* Artificial Narrow Intelligence (ANI)
AR, *see* Augmented reality (AR)
Artificial general intelligence (AGI), 277, 314, 329
Artificial intelligence (AI), 293
    accounting system, 371
    applications
        cognitive bias, 298
        collaborative partner, 295
        creativity *vs.* computation, 299
        environmental/fuzescanning, 294
        ethical boundaries and trust, 299
        fuzescan sources, 295
        global design/strategy team, 297
        real-time pattern recognition, 298
        scenario development, 298
        trend monitoring, 295
    benefits, 373
biodigital buildings, 264
blending design
    across sectors, 270
    bio-hybrid practices, 270
    design and emerging domains, 269
    designers, 269
    initiatives, 269
    transformative impacts, 268
climate disruption, 320–322
design thinking, 9
ethical implications
    accountability frameworks, 257
    accountability vacuums, 257
    bias issues, 257
    blend cutting edge, 258
    hallucination/generation problem, 256
    speculative scenarios, 257
    text-to-image models, 257
ethicists, 370
foundational challenge, 371
futuristic infrastructure, 315–317

Artificial intelligence (AI) (*cont.*)
   generative modeling, 371
   haunting risks, 315
   human-driven processes, 249
   interactive media
      design, 265–268
   interviews, 271–290
   key opportunities, 293
   metamodern cognitive
      framework, 370
   neurodiversity, 324–327
   philosophical and mythological
      lenses, 329–333
   policy making, 304–306
   post-human and virtual reality
      design, 313–315
   post-scarcity, 322, 323
   reflections and
      predictions, 368–370
   roads/highways, 200
   scenario planning/analysis, 250
      analyze scenarios, 252
      environmental scanning, 250
      frameworks, 251
      hybrid workflow, 252
      multiagent simulations, 251
      signal processing, 251
   societal system (*see*
      Societal system)
   speculative design
      automated curator, 254, 255
      future-scanning, 256
      intelligent sensor, 255
      Shaping Tomorrow, 253
      strategic planning/
         innovation pipelines, 254
   workshop models/projects
      background contexts, 300
      collaborative futures, 301
      critical issue, 303
      dialog partner, 301
      foresight methodologies, 302
      foresight sandboxes, 300
      research assistant, 302
      scenario generation, 303
      trend analysis, 303
      workshop model, 302
Artificial Narrow Intelligence
   (ANI), 277
Artificial Super Intelligence
   (ASI), 314
ASI, *see* Artificial Super
   Intelligence (ASI)
Augmented reality (AR), 200
Autonomous Mobility Network
   (AMN), 181

B

Backcasting method
   actionable pathways, 122–125
   approach, 115
   conceptual framing, 116–119
   definition, 108
   inspirational visions, 116
   path dependencies/incremental
      thinking, 115
   progress measurement, 124

roadmapping, 119–121
root cause analysis, 118
strategic foresight (*see* Strategic foresight and backcasting)
transitional roadmaps, 117
Backcasting system
external influences, 392
foresight method, 391
megatrends, 392
normative scenarios, 392
planning method, 391
plausible futures, 391
resilience, 393
sensemaking, 393
signals/wild cards, 394
systematic examination, 392
systems thinking/trend analysis, 393
uncertainties, 391
visioning/transformation, 394
zooming in/out, 394
Banking system, 353–355
BCI, *see* Brain–computer interface (BCI)
Biomimicry systems, 358
Brain–computer interface (BCI), 64, 313

C

Causal layered analysis (CLA), 388, 391
CLA, *see* Causal layered analysis (CLA)

Climate adaptation
design, 320–322
Comprehensive project
analysis/synthesis, 165, 166
autonomous mobility network, 172
backcasting, 143
backcasting and decision pathways, 138
blend techniques, 136
community-centric mobility platform, 173
environmental scanning, 137
external drivers, 151, 152
focal area selection, 139–141
focal issue/domain, 136
governance and review, 138
initial manual process, 147
mapping, 166
potential strategies, 177
presenting/peer reviewing, 139
prototyping, 138
prototyping techniques, 142
purpose statement, 146–148
reflection, 145
reimaginative solutions, 141
research, 137
review process, 139
RideNow, 144
ride-sharing service, 145
scenario building, 138, 142, 176–179
sensemaking, 137
signal collection, 159

INDEX

Comprehensive project (*cont.*)
    economic signals, 161
    environmental signals, 162
    figma, 159
    political signals, 163
    social signals, 160
    technological signals, 161
    values-based signals, 164
  solutions ecology, 144
  spanning paradigms, 142
  speculations/
    hypothesis, 166–171
  stakeholder and domain
    research, 137
  STEEP+V model, 152–159
  strategic foresight, 143
  strategic foresight
    analysis, 148–151
  strategies, 149, 171–176
  strategy selection
    community-centric mobility
      platform, 183
    decision-making
      process, 180
    decision matrix, 181
    design-futures practices, 186
    elements, 181
    logistics, 182
    strategic foresight playbook,
      184, 185
    sustainable mobility
      ecosystem, 183
    systemic risks/equity
      flags, 182
    trade-offs legible, 183
    trigger thresholds, 182
    wind-tunneling, 181
  sustainable mobility
    ecosystem, 175
Contemporary design thinking,
  (*see* Design thinking)
Contextualization
  culture/history/place-based
    identities, 130
  frameworks/tools, 134–136
  holistic perspectivism, 133
  indigenous communities, 130
  place-based intelligence, 131
  quantitative data and
    analytics, 135
  role of place, 132–134
  storytelling artifacts, 135
Cultural/social system, 350–352

D

Dator, Jim, 236
Design thinking
  AI technologies, 9
  content, 5
  evolution of, 4
  futures thinking, 9
  history timeline, 3, 4
  mechanization methods, 4
  movements/figures, 6
    Art Nouveau, 6
    arts and crafts, 6
    Bahaus, 6

modernism, 7
natural materials, 7
ornamentations/sleek designs, 7
postmodernism, 8
sustainability, 8
prehistoric landscape, 3
toolkit, 10
Development/analysis scenarios
cognitive biases, 76
coherent narrative description, 76
electric vehicle (EV), 78
implications analysis, 79
key phases, 76
methods, 77
strategy adaptation, 79
Development and analysis scenarios, 76
cognitive biases, 75
comprehensive description, 90
Dator, Jim, 87–89
food systems scenario, 93
futures wheel/cone, 81–86
    clusters, 85
    high-level framework, 82
    impacts/dynamics, 84
    map impacts, 82
    outer edges, 84
    scenario development, 81
    scenarios, 86
    sectors and dimensions, 82
    states and dynamics, 82
    techniques, 83

human–machine symbiosis, 90
world-building process, 89–93

E

Economic theories, 323
Environmental/ecological system, 348–350
Ethical/sustainable design
biodiversity outcomes, 46
biotechnological capabilities, 44
conceptualization, 45
core-philosophy, 46
encounters, 49
human and non-human life, 44
long-term outcomes, 48, 49
multifaceted complexity, 43
steering, 47
sustainability, 45
whole-system foresight, 49

F

Family/relationship system, 362–364
FAR, *see* Field Anomaly Relaxation (FAR)
Field Anomaly Relaxation (FAR), 386
Financial/banking system, 353–355
Food system, 360–362
Foundational methodologies
anticipating process
    congruous design, 71
    futures cone, 65–68

INDEX

Foundational methodologies (*cont.*)
    merging artifacts, 72
    principles, 61
    prototyping future, 70
    scenarios, 71
    signal recognition process, 68–70
    speculative design, 70–72
    STEEP+V model, 64–66
    systematic monitoring, 61, 62
    systematic process, 61
  approaches, 53
  capabilities, 55
  classical theories, 59
  concepts/approaches, 54, 55
  data-driven modeling, 56
  human and machine intelligence, 56
  influential futures, 60
  sense-making practices, 60
  speculative design
    data-driven analysis, 57
    design fiction, 58
    provocative artifacts, 59
    scenarios and contexts, 57
    Superflux, 58
    systematic process, 57
    transformative visions, 59
  theoretical foundations/paradigms, 59, 60
Future mobility integration, 227
  accessibility, 228
  archetypes, 237
  autonomous vehicle dominance, 227
  cultural shifts, 232
  data management, 229
  distance and living patterns, 233
  economic globalization effects, 232
  economic shifts, 229
  economy, 236
  education system, 231, 234
  energy sources, 229
  environmental impacts, 230
  futures wheel, 232
  global connectivity, 231, 235
  healthcare delivery, 230
  healthcare system, 233
  legal and ethical landscapes, 234
  legal/ethical considerations, 231
  multimodal transportation, 227
  personalized mobility services, 228
  psychological and cultural shifts, 234
  psychological impacts, 231
  road usage patterns, 228
  social interaction, 230
  travel time and distance, 230
  urban air, 228
  urban and rural areas, 235
  urban planning revolution, 229
Futures thinking
  approaches, 54
  artificial intelligence (AI), 249–259

business and management, 18
decision making
    strategies, 24
designers/technologists, 23
design thinking, 9
education and learning, 19
environmental and
    socioeconomic trends, 16
ethical and sustainable
    design, 43–49
ethical considerations, 10
evolutionists, 16
foundational
    methodologies, 53–72
integration, 17–20
intuitionists, 15
participatory methods, 24
perspectives/cultural
    impact, 33–42
preferable trajectories, 25
probabilistic forecasting
    models, 15
prospective, 15
scenario planning/speculative
    design, 25–28
strategic foresight, 29
studies, 15–17
sustainable development, 19
synthesis methods, 188
systems-oriented approach, 24
systems theory, 16
technology governance, 18
theorists, 15, 16
traditional practices, 23
typology (PPPP), 25
urban planning, 19

G

Generative methods, 321
Goertzel, Ben, 369

H

HCI, *see* Human–computer
    interaction (HCI)
Human–computer interaction
    (HCI), 5

I, J, K, L

Intelligent transportation
    systems (ITS)
    cybersecurity challenges, 210
    data generation and
        collection, 208
    driver behavior, 209
    economic impacts, 209
    educational system
        adaptations, 211
    education and career paths, 215
    energy consumption, 208
    environmental effects, 210
    global and economic shifts, 217
    global competitiveness, 212
    healthcare impacts, 211
    health landscape, 214
    infrastructure requirements, 209

INDEX

Intelligent transportation
     systems (ITS) (*cont.*)
   insurance industry
       disruption, 210
   legal and ethical
       considerations, 211
   legal and ethical landscape, 216
   psychological adaptation, 216
   psychological effects, 212
   public transportation, 210
   redefined urban
       landscapes, 213
   road safety, 208
   social dynamics, 215
   societal shifts, 211
   traffic flow, 208
   urban planning
       transformation, 209
   work-life patterns, 214
Interactive media design
   cognitive multiplier, 266
   disciplines, 263
   encompassing disciplines, 265
   implementation unity
       sentis, 265
   interactive digital media, 268
   responsiveness, 268
   rigorous principles, 268
   storytellers model, 266
   UI/UX design, 266
Interviews
   academic and professional
       circles, 283
   accountability, 272
   circular business models, 283
   collaboration, 277
   concerns/excites, 272
   deep reflections, 271
   demographic shifts, 279
   design emergency
       project, 273
   diverse multidisciplinary
       teams, 287
   economic and digital, 279
   emerging technologies,
       277, 284
   emerging trends, 283
   factors, 276
   foresight techniques, 284
   futures research, 281
   futurists/foresighters, 290
   governance perspective, 278
   governments/international
       organizations, 279
   innovators and
       entrepreneurs, 289
   integration, 280
   operations and reduction, 285
   primary quality, 285
   scenario planning, 286
   simplicity, 271
   social/political/ecological
       challenges, 274
   strategic foresight, 282
   technology, 272
   uncertainty and volatility, 276
ITS, *see* Intelligent transportation
   systems (ITS)

M

Machine learning (ML)
  computational fuzescanning, 293
  interactive media design, 263
  open-ended cocreation
    loops, 265
Maeda, John, 271–290
Metal-organic frameworks
  (MOFs), 321
ML, *see* Machine learning (ML)
MOFs, *see* Metal-organic
  frameworks (MOFs)

N

Natural language processing
  (NLP), 250, 253
Neurodiversity
  concepts, 325
  developmental trajectories, 327
  dysarthria, 324
  inclusion, 324
  operationalize, 325
  psychosyndemics, 326
  traditional fields, 326
NLP, *see* Natural language
  processing (NLP)
Non-player characters (NPCs), 266
NPCs, *see* Non-player
  characters (NPCs)

O

Olynick, Diana, 271–290

P, Q

Perspectives/cultural impact
  assumptions, 35
  case studies, 37
  diverse viewpoints, 33
  evaluation criteria, 34
  experiences, 36
  multifaceted/pluralistic
    design, 35–37
  pluriversal futures, 36
  Strategic Foresight Studio, 37–42
  treat European perspectives, 35
Philosophical/mythological lenses
  history, 329
  mental modeling, 331–333
  problem framing/research
    phases, 333
  speculative design, 332
  transdisciplinary design, 332
Policy making (AI)
  fundamental level, 304
  objectives, 304
  scanning and modeling, 305
  simulation sandboxes, 305, 306
Post-human and virtual reality
  design, 313–315
Post-scarcity economy, 323, 324
Prototyping techniques
  approaches and techniques, 94
  artifacts, 101
  community dialogue, 95
  concepts/models, 102–104
  diverse knowledge streams, 103

INDEX

Prototyping techniques (*cont.*)
   locations/environments, 100
   multimedia persona, 100–104
   path dependency and
      reactivity, 95
   radical reimagining/
      innovation, 96–99
   redefine premises, 96
   scenario planning, 93
   shapes XR prototyping tool, 102
   speculative societal
      artifacts, 103
   systemic prototyping, 102

R

Relationship system, 362–364
Resources, 395–399
Roadmapping process, 120–122

S

Scenario planning process, 26
Societal system
   biometric agricultural
      models, 358–360
   cultural/social system, 350–352
   environmental/ecological
      system, 348–350
   family/relationship
      system, 362–364
   financial/banking
      system, 353–355
   food/agriculture, 359–361
   futures approaches, 337
   health management, 338–340
   holistic systems mapping, 347
   infrastructure, 356–358
   law/justice/political systems
      addressing issues, 342
      capabilities, 341
      crime/legal issue, 343
      dynamic governance
         models, 341
      judicial system, 344
      politicians and legal
         practitioners, 345
      speculative artifacts, 344
      swarm intelligence, 343
      techniques, 342
      volitional governance, 342
   media/information
      system, 345
   transportation, 355–357
SOFI, *see* State of the Future
   Index (SOFI)
Space exploration, 312, 313
Speculative design, 29–31
   artifacts, 191
   digicars, 188
   direct platforms, 190
   exploration, 188
   gather initial insights, 192
   insight gathering, 190
   machine learning models, 190
   social exploration/strategic
      planning, 189
   version/build numbers, 191

State of the Future Index (SOFI), 389
STEEP+V (Social, Technological, Economic, Environmental, Political+Values), 64–66
  accuracy workflow
    data and classification, 194
    image generation, 195
    key phrases, 193
    labels, 192
    Midjourney, 194, 195
    signal card creation, 196
    signals categorization, 197
  analysis, 157
  approaches, 153
  concrete narratives, 156
  economic signals, 155
  environmental signals, 155
  external drivers, 151, 152
  key signals, 153
  political signals, 155
  RideNow, 153
  ride-sharing industry, 153, 154
  roads/highways
    aging population, 198
    AI and sensor data, 200
    augmented reality (AR), 200
    autonomous vehicle (AV), 199
    categorization, 207
    economic phases, 201, 202
    electric vehicle (EV), 199
    environmental impacts, 202, 203
    intelligent transportation systems, 209
    materials, 200
    political issues, 204, 205
    prominent patterns, 207
    public health considerations, 198
    safety innovations, 199
    social impact, 198
    technological impact, 199
    urbanization and smart cities, 198
    values, 205–207
  social signals, 155
  SWOT internal diagnostic RideNow, 158
  technological signals, 155
  values-based signals, 156
Strategic foresight and backcasting
  approaches, 109
  brands and product portfolios, 113–115
  definition, 107
  inner practice, 111
  innovation pipeline, 110
  investment decisions, 109
  Microsoft Office Labs, 110
  scenarios and dynamic simulations, 112
  short-term quarterlies/operations, 109
  techniques, 108
  technological and market discontinuities, 111–113
  trend impact analysis, 112
Strategic foresight design, 29, 30

INDEX

Strategic Foresight Studio
    climate change, 40
    criteria and instruments, 40
    design fiction, 38
    diverse futures
        methodologies, 37
    embedding futures methods, 42
    evaluation, 41
    experiential futures, 39
    external reviewers, 40
    firms, 42
    healthcare, 39
    learners report, 40
    methodologies, 38
    method/quality, 40
    primary objectives, 37
    scenario planning, 38
    student projects, 39
    Urban future, 39
Sustainable/resilience
        infrastructure, 217
    biodiversity, 221
    climate-related
        infrastructure, 220
    climate resilience, 218
    construction practices, 218
    cultural norms and values, 226
    economic shifts, 219
    educational system
        adaptations, 221
    financial products, 222
    futures wheel, 223
    global competitiveness
        realignment, 221
    global politics, 225
    global resource demands, 220
    health improvements, 226
    innovative financial
        systems, 225
    material technologies, 218
    natural systems, 224
    policy and regulation
        changes, 219
    psychological and health
        effects, 222
    public health
        improvements, 219
    redefined concept, 227
    reduced environmental
        impact, 217
    renewable energy, 218
    reshape learning, 225
    resilient infrastructure, 224
    societal values, 221
    transportation modes, 220
    urban and rural areas, 220, 223
    urban planning
        transformations, 219
    water management, 218
Synthesis methods
    comprehensive (*see*
        Comprehensive project)
    contextualization, 130–136
    democratic social media, 128
    evidence base, 129
    frameworks, 127
    future mobility
        integration, 227–237

intelligent transportation systems, 213–217
method transparency, 129
scope/evidence notes, 128
speculative design, 188–192
sustainable and resilience infrastructure, 217–227

T, U, V

Tool box software, 375
   causal layered analysis (CLA), 388
   chaos and nonlinear dynamics, 387
   cross-impact analysis, 376
   decision-making, 383
   decision modeling, 380
   delphi method, 376
   environmental scanning, 375
   field anomaly relaxation, 386
   futures wheel, 376
   genius forecasting/intuition/vision, 385
   heuristics modeling, 388
   interactive scenarios, 383
   morphological analysis, 377, 381
   normative forecasting, 386
   participatory methods, 384
   personal futures, 389
   perspective concept, 387
   planning scenario, 383
   polygon, 378
   prediction markets, 385
   relevance trees, 382
   scenarios, 377, 382
   simulation and games, 384
   SOFI software system, 390
   state of the future index (SOFI), 389
   statistical modeling, 381
   S&T road mapping, 386
   structural analysis, 379
   substitution analysis, 380
   systems perspectives, 379
   technology sequence analysis, 381
   trend impact analysis, 378
   vision, 385
   wild cards, 379
Transportation/infrastructure
   collapse scenario, 244
   continuity scenario, 243
   discipline scenario, 244
   economic crises, 238
   mobility revolution, 240
   potential implications, 242
   scenarios, 241, 242
   strict regulations, 239
   technology and mobility, 237
   3D engine, 242
   transformation scenario, 245
Transportation system, 355–357

W, X, Y, Z

World-building techniques, 91–95